BAD CHOICES

HOW Algorithms can Help You Think smarter and Live Happier

ALI ALMOSSAWI

JOHN MURRAY

First published in Great Britain in 2017 by John Murray (Publishers)
An Hachette UK Company

1

© Ali Almossawi 2017

The right of Ali Almossawi to be identified as the Author of the Work has been
asserted by him in accordance with the Copyright, Designs and Patents Act 1988.

Creative and art direction by Ali Almossawi
Illustrations by Alejandro Giraldo
Design by Spring Hoteling

A CIP catalogue record for this title is available from the British Library

Hardback ISBN 978-1-47365-076-3
Trade paperback ISBN 978-1-47365-077-0
Ebook ISBN 978-1-47365-075-6

Typeset in Palatino

Printed and bound by Clays Ltd, St Ives plc

John Murray policy is to use papers that are natural, renewable and
recyclable products and made from wood grown in sustainable forests.
The logging and manufacturing processes are expected to conform to
the environmental regulations of the country of origin.

John Murray (Publishers)
Carmelite House
50 Victoria Embankment
London EC4Y 0DZ

www.johnmurray.co.uk

For Fatima

YOU CAN SWIM ALL DAY
IN THE SEA OF KNOWLEDGE
AND NOT GET WET.

—Norton Juster, *The Phantom Tollbooth*

PREFACE

Did you know that Richard Feynman started developing the equations that won him the Nobel Prize after seeing someone throw a plate in the air? Or that John von Neumann modeled parts of his electronic computer on a friend's idea about how memories are stored in the human brain? Or that the sight of a kicking and screaming orangutan at the zoo led Charles Darwin to his big idea? What Feynman, von Neumann, Darwin, and others have in common is that they see physics and mathematics and science everywhere, way beyond the confines of their laboratories.

Even if you're not gunning for a Nobel Prize, you probably do things in your everyday life that can be modeled as algorithms. In fact, you apply them on a daily basis to solve various problems: finding pairs of socks in a pile of clothes, deciding when to go to the grocery store, determining how to prioritize your tasks for the day, and so on. An algorithm is a series of unambiguous steps that achieves some meaningful objective in finite time. The series might begin with some input and is expected to produce an output. Those are an algorithm's characteristics. What's fascinating is that Babylonian tablets from around 1800 to 1600 BCE reveal that ancient Babylonians wrote down their procedures for determining things like, say, compound interest or the width and length of a cistern given its height and volume using algorithms. That is to say, their procedures were made up of an unambiguous series of steps; they had some input, some output, they eventually terminated and they were useful. Algorithms can thus be found in the works of various contributors to mathematics over the centuries. After the advent of computers, these characteristics have proved crucial because they allow computers to carry out tasks in a way that is predictable.

Despite the importance of algorithms in our lives, texts on the subject tend to focus largely on intricate details—the "how"—while perhaps ignoring the more practical lessons of those algorithms that make them appealing. The seemingly simple everyday tasks we just mentioned can be undertaken in a number of different ways. The more aware we are of those ways, the better we can hone our ability to achieve a task in the most efficient way. Think of it like enhancing a general-purpose intuition that we all possess. That's where *Bad Choices* comes in. This book aims to

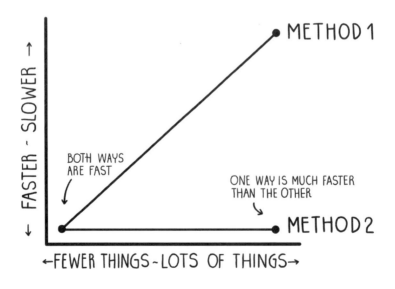

acquaint you with algorithmic thinking by highlighting different ways of approaching everyday tasks and pointing out how these approaches fare *relative* to each other. For instance, two methods of looking for a shirt in your size on a rack of shirts might be described like the graph on the opposite page.[*]

Those shapes of lines have names like *linear* and *logarithmic*, which we will flesh out and discuss throughout the book. And while both approaches are comparable in terms of performance when we have a few things, notice how that changes as the number of things increases. This book includes twelve familiar scenes, such as a living room, a tailor shop, and a department store. In each scene, there are a number of potential tasks to be done. After each illustration, a paragraph describes the scene, and a few pages of commentary and discussion relate the scene to concrete concepts from computer science and highlight at least two possible ways of undertaking the fundamental task at hand. One that's slower and one that's faster. That difference is what the book's title aims to emphasize, albeit somewhat provocatively. The title is partly inspired by computer scientist Donald Knuth's talk of "good" algorithms, which is to say fast or effective ones.[†]

---

[*]   Note that all the lines in this book are plotted on a log-log scale, which is why the shapes of lines are the way they are.

[†]   It's important to mention at the outset that these qualifiers don't necessarily generalize to other things in life, like learning, where speed should not be considered a virtue. In my experience, any learning environment that encourages students to be fast learners is setting its students up for failure.

BAD CHOICES

INTRODUCTION

Why Focus on Relative Magnitudes?

Comparisons are amazingly powerful. One of the first things children learn are abstractions like *big* and *small*, which is why when a child asks, "How tall is that titanosaur that they now have at the Natural History Museum?" one finds that it is less meaningful for the child to hear the response, "Seventeen feet tall, little one," and more meaningful to hear, "If Ms. Susan, Ms. Margaret, and Mr. Jascha were to stand on each other's shoulders, Mr. Jascha would probably be able to tickle the titanosaur's chin."

Thinking in terms of relative magnitudes may in fact be an ability that we are all born with. Recent experiments seem to suggest that babies show as much brain activity in response to a change in an image they are seeing as they do in response to noticeable changes in the *number* of images they are seeing. Other experiments in remoter parts of the world suggest that people who have not been subjected to what we might call formal education reason about numbers in terms of orders of magnitude. It's an intuition that we appear to have innately.

One subspecies of humans whose appreciation for this intuition is manifest is computer scientists. It's what gives them the ability to recognize, fairly quickly, which of several competing approaches to solving a problem might be better. That fact is a reminder that seeing things in terms of relative magnitudes is an ability that remains useful even after you develop mastery of a field. Think of it like the mathematical notation you learned in primary school, which you continued to use throughout school and college and beyond.

This idea is a large part of the motivation behind writing this book. I had long used comparisons, estimates, and approximations to understand various concepts during my school and college years, but I dared not admit that to anyone because it felt like a less sophisticated way of learning. It wasn't until I read books like *The Strangest Man*[*] and *The Society of Mind* that I realized I wasn't the only one who found that way of thinking useful. Much later, I read *The Art of Insight in Science and Engineering* and similar books, which talk of the same idea and its implications for insight.

It is my hope that this book impresses on you the ability to better think about decisions throughout your life and better understand what tradeoffs they come with. The book doesn't aim to teach you how to better match socks, an intuition that most people will likely already have, but rather to persuade you to turn the mirror on yourself and ask, "I didn't realize I could think about *my* socks in that way." Much like critical thinking, algorithmic thinking is a highly capable tool that has the potential to impact behavior for the better.

---

[*] There's a funny passage in that book about one Oliver Heaviside, "an acid-tongued recluse" whose approach to teaching the mathematics of engineering was a pragmatic one. "Engineers prized Heaviside's methods for their usefulness, but mathematicians mocked them for their lack of rigour. Heaviside had no time for pedantry ('Shall I refuse my dinner because I do not understand digestion?')."

Why Focus on Everyday Tasks?

Algorithms can be complex, but they're also critical and often already a part of our lives. We just don't know it or think about it much. By highlighting those parts of our lives that serve as good models for various algorithms, we end up with an approach that has several benefits.

It is relatable: Many of the explanations in this book leverage illustrations. Explaining by way of illustrations is not only useful because of the appeal that illustrations bring to an otherwise pedestrian work, but also because illustrations can place one in a world that is relatable, and relatable worlds are engaging and encourage you to be more sophisticated in your reasoning as you connect newly acquired knowledge with what you already know. That is precisely why analogies are so effective.

It is interactive: If you look at human history, you're likely to find that many of the names you recognize belong to people who were educated as apprentices rather than as note takers. Algorithms are often referred to as recipes, but I find following recipes to be too much like note taking: dry, rote, and vapid. In such a model, you're seen as a sort of container and your instructor's task is to pour in the knowledge. To borrow yet another metaphor, it's like watching a sitcom with a laugh track, in which somebody else is doing the laughing on your behalf. In this book each lesson is presented through an everyday scenario, encouraging you to develop your own understanding by interacting with the scenes, talking through them, and thinking beyond them to your own life and daily routines. I believe this interactive approach makes for a more compelling read and a more engaging learning experience. My greatest memories of childhood learning involve conversations with one of my parents or with a teacher, all of whom seemed to realize that process is as important for learning as intelligence.

It acknowledges multiple outcomes: One of my favorite lines about learning comes from Francis Bacon, "That use which is collateral and intervenient is no less worthy than that which is principal and intended." Questions can have more than one possible answer. Works that are more exploratory are therefore more amenable to having multiple outcomes. One might imagine them like a science museum—parents walk up to an installation, read a label, and then try to explain to their children the lesson embodied in that installation. No one comes in a scientist, no one leaves a scientist, and yet everyone gets something valuable from the experience.

1.

MATCH THOSE SOCKS

Baroness Margie Wana is a member of a formerly influential Viennese family who was recently indicted for smuggling Kinder Surprise eggs into the United States. She now works as an au pair in Bern and is folding clothes for the first time. Margie is shocked to discover that each member of her host family sweats through a pair of socks every half hour, making finding and matching pairs more time-consuming than she ever imagined. On the plus side, they have different shoe sizes and like different colors.

Hint: While there might be several tasks here, perhaps start with the fundamental one.

Have you ever thought about how essential a biological feature *memory* is in humans? The image of someone leaning back into her chair, putting one hand to her forehead, and pressing her eyes shut as she calls to mind a verse or an equation or a telephone number—that's the quintessential human. Imagine the struggle of having to go through life without that feature, as do sufferers of dementia. For starters, you would end up having to repeat a lot of the same work. Like in *Memento*, where every morning the protagonist has to fill up his mind all over again with all the bits of information that he needs to carry out his primary task.

I mention this at the outset because of the fact that faster methods of solving problems are often faster because they happen to leverage memory.[*] Consider AlphaGo, which last year beat a champion at the game of Go thanks to its ability to learn not only from expert humans but also from itself, thus amassing a greater memory from which to work.[†] Put differently, many of the faster ways of solving problems that we will encounter in this book, simple though they are, are fast because of their ability to avoid doing the same action on the same thing multiple times.

Let's not get ahead of ourselves though. Back to the socks, and to poor old Margie Wana, the irony of whose name was lost on the federal agents who recently seized her stash of chocolate goodness. Margie is facing the daunting task of matching pairs of socks from a humongous pile of clothes. Let us focus on one of several tasks that exist here and consider two possible methods for taking on that task:

---

[*]   A fact that is sometimes characterized with the phrase, "trading memory for time."

[†]   This approach was pioneered at the University of Toronto a decade ago and is referred to as deep learning.

OBJECTIVE: MATCH THE PAIRS OF SOCKS IN THIS PILE OF CLOTHES.

METHOD 1: PICK A SOCK, LOOK FOR ITS MATCH IN THE PILE, PUT IT TO ONE SIDE. THEN PICK ANOTHER SOCK, LOOK FOR ITS MATCH IN THE PILE, AND PUT IT TO ONE SIDE. AND SO ON.

METHOD 2: PICK A SOCK, PUT IT TO ONE SIDE. PICK ANOTHER SOCK. IF IT MATCHES ANY OF THE ONES SHE HAS PUT TO ONE SIDE, MATCH IT. OTHERWISE, ADD IT TO THE LINE OF UNMATCHED SOCKS, LUMPING IT WITH SOCKS THAT HAVE THE SAME COLOR OR SIZE.*

---

* Note that with both methods, we ignore the task of separating socks from non-socks because we're focused on the fundamental task of matching socks.

Before reading any further, I would suggest working through these scenes using pen and paper, props, or whatever else you feel comfortable with. Think about what achieving the objective entails in terms of individual steps and assumptions. Try to do that for all the scenes that follow.

With a pile of, say, four pairs of socks, it doesn't really matter which method Margie uses—she will be done fairly quickly. Now imagine Margie with hundreds of socks in front of her. If she opts for the first method, the chance that she will come across the same old sock over and over again is quite high, since she never takes it out of the pile. When she first comes across it, she simply doesn't glean any information from it. With the second method, however, she keeps a line of unmatched socks to one side, thus ensuring that she only ever comes across a sock in the pile once. The second method, therefore, ends up being faster because of its reliance on memory. More precisely, because of what's sometimes called a *lookup table* or *cache*. Though it need not be, it is useful to think of a lookup table as a collection of unique identifiers (keys) each pointing to some associated item of data (values) where you, quite literally, look up the values of keys. We call this type of representation a *key-value pair*. In this case, our keys would likely be "color." When Margie comes across, say, a red sock, she looks up "red" in her line of unmatched socks. If she finds a "red" area, she might then look for additional identifiers like, say, style or hue and take things from there. Otherwise, she would create a new "red" area with the solitary red sock.

Here are how the two methods compare.* Notice that Method 1 becomes noticeably slower than Method 2 as the number of socks in the pile grows. There are many more ways of tackling the tasks in this book

---

* There are more nuanced ways of looking at these rates of growth. One is in terms of whether a particular method grows no faster than the rate shown (known as *big-o notation*) or no slower than the rate shown (known as *big-Ω notation*, and read *big-omega*). Another is to consider whether the rates of growth describe the best, worst, or average cases. We will be talking about these different cases later on.

than the two methods that are highlighted. The discussion is meant to emphasize two methods that are notably different in their asymptotic rates of growth, leaving out other methods whose performance may fall somewhere in between. With this scene, for instance, Margie might have chosen, alternatively, to find matches by way of the pigeonhole principle, which would have involved pulling six socks from the pile at a time and matching pairs that way.

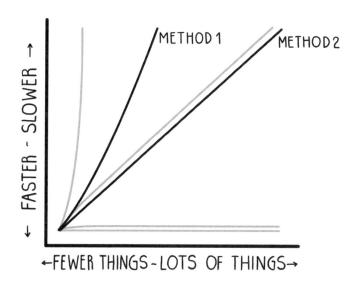

Now, when we pick out a sock from the pile, we could probably tell fairly quickly if we've come across its matching pair. Most people's short-term memory is quite good at remembering half a dozen or so groups of things, which is what we have here. And so, coming across a sock in the pile that we've already placed to one side ought to elicit a near-immediate "Ah, I've seen that one before!" If you've ever played a card-matching game like *Memory*, the powers and limitations of that faculty ought to be familiar to you.

If we did have a bigger disparity in sock types and colors, however, our line of unmatched socks might grow quite a bit, forcing us to scan through the whole line every time we pick something new out of the pile. Scanning through a line of things, an *array*, can be time-consuming when the number of things is quite large, reason being that the thing you're looking for might be at the very end of that line. We would thus have to pass through the entirety of the array first.

In 1953, and while working at IBM, the mathematician Hans Peter Luhn drafted an idea that gave way to an alternative structure to help alleviate the potential slowness that is inherent in array lookups. It's sometimes called an *associative array*, a *dictionary*, or a *hash table* (more salt in the wound for poor Margie). A hash table does precisely what an array does, in that it stores things in a collection, except it trades order (i.e., large black sock comes after small red sock) for near-immediate lookups, also referred to as *constant-time* lookups.[*]

They are called constant time because the lookup no longer depends on the length of the array. Rather, its speed is independent of the length of the array. This observation is usually true, though not always. A lot of things in software are like that, to the chagrin of many a researcher and practitioner. There are no fundamental laws in software like there are in nature. In this example, we're assuming that because of the small number of disparate socks, Margie's synapses will fire so quickly that her reaction will be near immediate.

As we will see later on, most constant-time lookups happen when we are able to model a task by way of a formula, which eliminates the need to step through a task and *iterate* over all the items at hand.[†] With hash tables, that formula is known as a *hash function*. Its job is to put an item into a pile and then perhaps retrieve that item at a later time fairly quickly.

All of that is just an aside, though. Our main takeaway from this scene is that an approach that reuses knowledge can be faster than one that doesn't. This is particularly useful to know when our task involves doing something over and over again—you're in a store, looking through

---

[*] In this example, Margie doesn't really care what order unmatched socks are in. All she cares about is that the socks are to one side. So the order is incidental. It is insignificant.

[†] For example, finding the sum of the first n numbers would be slower if you were to iterate over those n numbers one by one, summing pairs of numbers each time. It would be much faster if you were to use a formula instead: $n \times (n + 1) / 2$.

a box of letter-shaped candles for your daughter's birthday cake. Or you're about to wash your clothes and need to separate your whites from your colors and from your delicates. Or you're trying to win a game by coming up with the longest word from a set of jumbled-up letters, as they do on the British TV show *Countdown*, where contestants have thirty seconds to come up with the longest word they can make from the nine letters in front of them.

In each of those situations, ask yourself if the task at hand could be made faster by leveraging memory—yours or the world's. With our pile of socks, we exploited the fact that we had no more than five variations of socks by maintaining a line of unmatched socks. With our box of candles, we would pick out any of the four letters we need the moment we come across it, rather than searching for "L" and then for "U" and so on.

With the dirty clothes, we might keep them in three separate baskets to begin with rather than rummage through the clothes prior to a wash. And with the longest word situation, we might look for any word that we can form at first sight and then see if we're able to lengthen it by, say, conjugating it or making it plural. In such a case, our initial choice serves as a sort of prefix (hence, memory) to subsequent words. There is a fascinating tree-based structure called a *trie* that does exactly that. It exploits the fact that words or numbers share prefixes and uses that knowledge to make things like spell-checking and auto-completing words that you might enter into a search box much faster.

ISN'T IT INTERESTING, HOW THE MUNDANE CAN TURN
INTO SOMETHING SO ENGAGING WITH A SLIGHT SHIFT OF THE HEAD.

2.

FIND YOUR SIZE

It's the day after Christmas, and Eppy Toam, a nurse from Inverness, has started camping outside of a department store in preparation for this year's Boxing Day sale. Her shirt size is quite common and she wants to make sure she is the first one into the store so that she can grab all the shirts that are in her size. She is going to have to be fast. Things get pretty ugly. Last year, fifteen people were injured and the army had to intervene. How might Eppy increase her odds of grabbing the shirts she's after before everybody else?

Hint: Consider pushing the example to an absurd limit. What if the racks were as wide as the store?

If we are searching for an item in a collection of items, then surely, it must be the case that we face the possibility of having to look through all of those items before finding what we're looking for? In other words, mustn't it be the case that if we have one hundred items, we face the possibility of having to look through one hundred items, which is to say, taking *linear time*? Generally speaking, a linear function means that if it takes us a minute to find something among one hundred things, then we can expect to spend two minutes to find something in a pile of two hundred things. Ordinarily, yes. But there is a particular quality that a collection can have—namely, the quality of being sorted—that allows us to find an item in *logarithmic* time. In other words, in seven or so steps rather than 100. Recall that a logarithm is simply the inverse of an exponent. When writing computer programs, we assume the base of a logarithm to be two, and so the logarithm of 100 is log_2 100, which comes to around seven. That massive improvement you see when going from linear time to logarithmic time is why the logarithm is such an important concept, particularly when we talk about rates of growth. It is a concept that we will revisit often throughout the coming chapters.

Let us first describe how Eppy might go about descending on the store, her face likely painted in the colors of glory, her tartan shawl billowing behind, her battle cries shooting past her rattling teeth and clinging on to the store's walls, for posterity to marvel at. She has been psyching herself up all morning.

OBJECTIVE: FOR A GIVEN RACK, FIND THE SHIRT IN HER SIZE.

METHOD 1: FOR A GIVEN RACK, LOOK THROUGH THE SHIRTS FROM ONE END TO THE OTHER.

METHOD 2: FOR A GIVEN RACK, START LOOKING FOR THE CORRECT SHIRT SIZE SOMEWHERE NEAR THE MIDDLE. IF THE SHIRT IN THE MIDDLE IS LARGER THAN HER SIZE, MOVE TO THE LEFT. IF IT'S SMALLER THAN HER SIZE, MOVE TO THE RIGHT. AND SO ON.

Here are how the two methods compare. Notice that Method 1 becomes noticeably slower than Method 2 as the number of shirts on a rack grows.

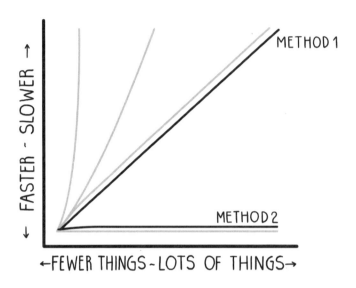

As you might have guessed, Method 2 leverages two pieces of knowledge. One, that the shirts are most likely sorted on the rack by size. And two, that because Eppy's size is a common size, which is to say, an average size, then it would likely be near the middle of the rack. Using that intuition not only to start from the middle, but to subsequently move left or right in jumps, thus halving the collection each time, is a signature of a *logarithmic-time* algorithm.* It is the same intuition we might use to look

---

* Similarly, the process of repeatedly doubling a number from 1 to n is also logarithmic in that we can make no more than $log\, n$ jumps before we get to n. For instance, how many years would it take to make a million dollars if we started with $1 and doubled it every year? We could do it by hand, or we could say $log_2\, 1{,}000{,}000 = 19.93$ years.

for a word in a dictionary or a name in a telephone directory or a topic in a book's index. The same intuition we would use if we fall asleep reading a tedious novel and want to pick up where we left off the next day. More generally, we might characterize this approach as being one of discarding information.

MS. EPPY FINDS HER SIZE
IN FOUR STEPS

MS. EPPY FINDS HER SIZE
IN TWO STEPS

For us, the most relevant thing about logarithms is that they grow slowly, as you saw in the previous graph. We like ways of solving problems that grow slowly, because it means our method is not as sensitive to the number of things that we're working with. In this scenario, Eppy would likely find her shirt among a rack of one hundred shirts in under seven steps, and among a hypothetical rack of one thousand shirts in only ten or so steps, which isn't too bad. This method of logarithmically searching for something in a sorted collection is sometimes referred to as *binary search*. It provides a great improvement over the alternative method (Method 1), known as *linear search,* and is certain to reward Eppy with a bagful of shirts.

3.

POP TO THE SHOPS

Iain Patoys is a retired linguistics teacher who lives in East London. He has a bad back from a fall a few years ago. He hates leaving the house because the neighbor's dog scares him, but alas, if he doesn't want to die of hunger, he has to pop to the shops every so often to buy his own groceries. This being London, it rains a lot, and Iain being Iain, doesn't like getting wet. How might Iain minimize the number of times he goes to the store within a given week without dying of hunger?

There is a timeless sketch that the British double act The Two Ronnies did during their many years on air.* In it, a customer walks into a hardware store and reads off the items on his list, one by one. Rather than waiting for his customer to read the entire list first, the store owner instead retrieves an item each time the customer reads it out, which ends up causing the store owner great distress.

Keep this vignette in mind. We'll come back to it in a bit. But first, let's see how Iain might go about deciding how often to go to the store.

---

* You can watch it at bookofbadchoices.com/links/ronnies. The relevant series of events happens near the middle of the clip.

OBJECTIVE: MAKE AS FEW TRIPS TO THE GROCERY STORE IN A GIVEN WEEK AS POSSIBLE.

METHOD 1: REALIZE HE HAS RUN OUT OF SOMETHING. HEAD TO THE STORE TO BUY IT.

METHOD 2: MAINTAIN A LIST OF THINGS THAT HE HAS RUN OUT OF. GO TO THE STORE ONCE THE LIST REACHES A CERTAIN SIZE OR ONCE HE RUNS OUT OF AN ESSENTIAL LIFE-AFFIRMING FOOD ITEM, LIKE KIT KAT BARS.*

---

*   Other product-placement opportunities are available.

Here is a graph we are already familiar with that gives us a visual sense of how those two methods compare.

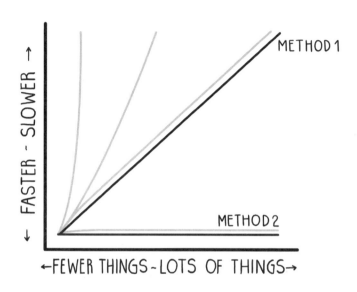

One interpretation of this scene would be to say that it's essentially about avoiding repetitive work, much like why a secretary tasked with hole-punching papers from ten different reports might choose to hole-punch them all in one go rather than one at a time. Or why you might scrub all your dishes first and then wash them, rather than scrub and rinse each one at a time. Or why you might cut an onion lengthwise before cutting it widthwise to dice it. Or why elevators in newer high-rise buildings have so-called destination dispatch systems that put passengers going to the same floors in the same elevator. There is another, subtler observation one could make, and that's to do with what triggers Iain's visits to the grocery store. Let's explore that for a bit.

In computing, there are lots of ways that collections of items are stored. We saw the most basic one earlier on, that of an *array* of unmatched socks. We then saw in the second scene how an array could maximize a particular quality, namely, searchability, by virtue of having its contents sorted. Recall the sorted shirts on the racks. That is essentially what these *data structures*, or *abstract data types* as they are sometimes called, do. They maximize one or more qualities that we might care about, usually at the expense of other qualities that we don't care as much about. More generally, we might say that some qualities are simply antithetical to each other. For instance, security and usability—an application that asks you for a password every time you click a button might be more secure than one that doesn't, but it's also less usable.

The structure that I think is relevant to us here is known as the *stack*. Similar to the metaphor its name invokes, a stack maximizes the quality of us only caring about the item that's on top, irrespective of how many things are beneath that top item. So you might go into a coffee shop and see a stack of newspapers, and you would only glance at the one on top because you know it to be today's paper, and you only care about today's headlines. And so with stacks, we care a lot about *peeking* at the top item.[*]

In Iain's case, his cognitive stack contains items that he has run out of, and he only concerns himself with going to the store and *popping* items off that stack, which is to say, removing the topmost one repeatedly until the stack is empty, whenever he *pushes* a Kit Kat onto the stack. In other words, the Kit Kat is the trigger for emptying the stack. Until then, he can happily bury all the other items in there and go about his business. And so, our callback to The Two Ronnies sketch is that there too, allowing the store owner to build a cognitive stack, perhaps one for each row

---

[*] These are the actual names of a stack's operations. As you might imagine, all these operations take a constant amount of time.

of shelves so that he doesn't have to go up and down the ladder multiple times, would have caused the store owner much less distress. The customer would read out his entire list of items, the store owner would build his stacks, following which he would walk the length of each row of shelves, popping the items for that row's stack, and so on.

In 1946, Alan Turing published a report introducing the concept of a stack using the term "burying." As Andrew Hodges notes in his biography of Turing, the idea came as news to von Neumann. Here is a short excerpt from Turing's report:

> How is the burying and disinterring of the note to be done? There are of course many ways. One is to keep a list of these notes in one or more standard size delay lines (1024), with the most recent last. The position of the most recent of these will be kept in a fixed TS [Temporary Storage], and this reference will be modified every time a subsidiary is started or finished.

It's humbling reading about how concepts that we consider intuitive nowadays came to be. It makes us realize that those concepts only became obvious after someone went through the trouble of spelling them out. For an interesting take on that, you might want to read about the Flynn effect, named after Jim Flynn, which suggests that humanity has been getting smarter in part because our intuitions continue to mature and become more sophisticated. New humans now come preinstalled with intuitions that are better than those of their forefathers. Reading old texts and treatises is always fun, as they show us how far we have come. I recall once picking up Desiderius Erasmus's *Handbook on Good Manners for Children*, published in 1530, in which readers are given advice such as "Don't let your nostrils be full of snot like some grubby person; Socrates was criticized for that vice." To a reader in the twenty-first century, the advice seems self-evident. And yet within the context of that period, it was novel.

Turing's talk of subsidiary operations brings to mind yet another real-life example where a stack would be useful. Imagine if the mailman were to stop by the next morning at Iain's house and refuse to look him in the eyes. A solitary tear runs down the mailman's cheek. His lips might perhaps quiver.

"Sorry, but did I do something to upset you?"

"As a matter of fact you did. As a matter of fact. You did," the mailman replies, swinging his gaze to the horizon.

The way in which Iain would try to recall how he might have offended the mailman would likely be akin to popping items out of a stack. Namely, his "mailman" stack. The metaphor is apt because his most recent interaction with the mailman is more likely to be the culprit than

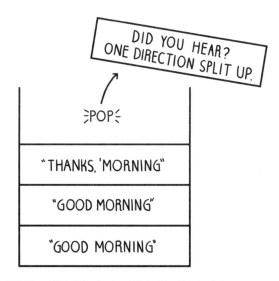

INTERACTIONS WITH MAILMAN

the previous one, and the previous one is more likely to be the culprit than the one before that.

Are there other things in everyday life that work like stacks? How about the Web? Every time you click on a link, you're pushing a Web site onto a stack, and every time you go back to a previous link, you're popping a Web site from that stack. You might not care how many Web sites you've visited in total, so long as you can go back to the last one, and from that one to the one before it, and so on.

For Iain, one hopes that he will be able to appeal to his cognitive stacks to make amends with his mailman and to better determine when to pop to the grocery store.

4.

BRING HIM HOME

Ioannis is lost in his own tailor shop. Iconic though his shop is among Athenians for its impossibly speedy turnarounds, Ioannis is known as a hoarder. A disposition made worse by the fact that his shop sits on a massive plot of land, most of which is unused. The shop has been expanding onto that land over the past three decades as Ioannis builds more passages and puts up more shelves to fuel his habit. And now here he is, somewhere in that messy maze, with walls covered in an endless array of thread rolls and clothes and broken sewing machines. How on earth is Ioannis going to find his way back to the front? Or is he destined to perish here in this labyrinth of his own creation?

So goes the Greek tale, that when the Minotaur was born—a creature half bull, half man—the great architect Daedalus built a labyrinth in which to place the ferocious being.

"Once inside, one would go endlessly along its twisting paths without ever finding the exit. To this place the young Athenians were each time taken and left to the Minotaur. There was no possible way to escape."

Luckily for Theseus, who was to be fed to the Minotaur, the king's daughter Ariadne fell in love with him and devised a plan to help him escape.

"She sent for Daedalus and told him he must show her a way to get out of the Labyrinth, and she sent for Theseus and told him she would bring about his escape if he would promise to take her back to Athens and marry her. She gave him the clue she had got from Daedalus, a ball of thread which he was to fasten at one end to the inside of the door and unwind as he went on. This he did and, certain that he could retrace his steps whenever he chose, he walked boldly into the maze looking for the Minotaur. He came upon him asleep and fell upon him, pinning him to the ground; and with his fists he battered the monster to death."[*]

Keep this vignette in mind. We'll come back to it in a second. First, let us describe the three methods that Ioannis might opt for to get back to the front of the store.

---

[*] Best followed with, "Nighty-night, kids."

OBJECTIVE: GET TO THE FRONT OF THE STORE.

METHOD 1: WALK THROUGH THE PASSAGES, MAKING RANDOM TURNS UNTIL HE FINDS HIS WAY OUT.

METHOD 2: HOLD HIS RIGHT HAND TO THE WALL AND FOLLOW THE WALL, ONLY MOVING TO THE RIGHT THE WHOLE TIME.

METHOD 3: TAKE ONE OF THE THREAD ROLLS OFF THE WALL AND UNWIND IT AS HE MOVES THROUGH THE PASSAGE. IF HE COMES TO A DEAD END OR IF HE STUMBLES ACROSS AN UNWOUND BIT OF THREAD, HE TURNS AROUND, WALKS TO THE LAST TURN THAT HE TOOK, AND TAKES A DIFFERENT PATH.

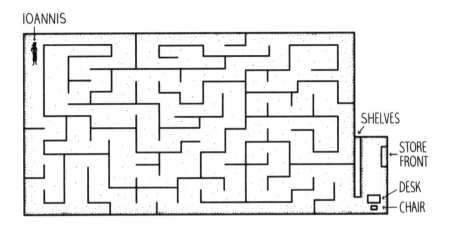

Method 1 mimics how a mouse might navigate its way around a maze. No advanced cognitive faculties, just a random walk from one place to the other until, by chance, it comes across a piece of cheese. In fact, this method is sometimes referred to as *random mouse*. As you might imagine, the approach can be slow.

Method 2 is a bit more interesting, albeit still amazingly simple. Here, Ioannis follows the wall with one hand, and by doing so, he eventually finds his way to the front of the store. Why does this approach work? It works because if you rearrange a maze's walls, you might in fact end up with a straight line. And so you can think of a maze as being like a piece of string, where it's quite clear that walking from one end of the string will eventually get you to the other end.[*]

Though faster than Method 1, the main problem with this approach is that it doesn't quite work if the maze happens to have so-called islands or loops in it. These are walls inside a maze that are not connected, either

---

[*] The illustration that follows is inspired by one of Jamis Buck's illustrations. See the references at the end of this book for further details.

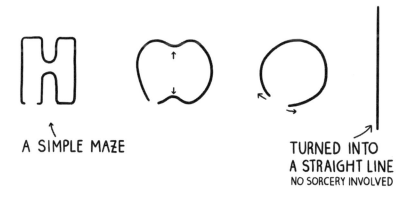

A SIMPLE MAZE

TURNED INTO
A STRAIGHT LINE
NO SORCERY INVOLVED

directly or indirectly, to the outer wall. In the 1820s, the fourth Earl Stanhope, a mathematician, built at Chevening in Kent the very first garden maze that had such loops in it. It was based on a design by the second Earl Stanhope, and its aim was to create a maze that could not be solved using approaches like Method 2. A loop in a maze might look something like the maze to the left.

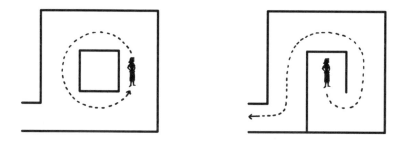

And so, with such mazes, this particular method of escaping, known as *wall follower* or the *right-hand rule,*[*] may fail to achieve the desired result.

---

[*] Alternatively, the *left-hand rule* if one chooses to follow the wall with one's left hand.

A quick aside. One of the most interesting facts about Charles Darwin is that he spent close to two decades looking into all the possible counter-arguments to his theory of evolution before he published *On the Origin of Species*. I've always imagined walking through a theory of that scale to be like walking through a maze, where forks along a path correspond to propositions,[*] where dead ends are realizations that an argument is invalid, and where exits are realizations that an argument's conclusion seems to logically follow from its premises. It turns out that this cognitive approach of venturing to the ends of arguments and counterarguments is equally applicable when trying to escape an actual maze. And that's precisely what we see with Method 3 and in the story about Ariadne's thread.

With this method, what Ioannis does is he makes use of *backtracking*, which is what the thread allows him to do—to backtrack whenever he hits a dead end and try a different path.[†] The ability to go back to intersections and try more promising routes guarantees that Ioannis will eventually find his way out. This strategy for escaping a maze is known as *Trémaux's algorithm*, an attribution that comes from French mathematician Edouard Lucas's *Récréations Mathématiques*, published in 1882. Recent research also seems to suggest that other creatures, namely ants, may use a form of backtracking to recover from a broken trail. The approach is faster than Method 1 and is able to handle loops, unlike Method 2.

Note that the three methods we've covered here are useful for when we are already inside a maze and want to get out. There are other methods to escape a maze that might be much faster but require that we know beforehand what the maze looks like. The methods we've seen here don't guarantee a shortest path out of the maze. Those other ones might.

---

[*] An argument is made up of a set of statements known as propositions, each of which is either true or false.

[†] He could have achieved the same thing by marking intersections with a piece of chalk or by dropping bits of cloth.

Now, you don't have to be lost in Rome's catacombs or a hapless character in a Stephen King novel for these maze-solving approaches to come in handy. There are dozens of real-life mazes all across the world, some spanning several miles, where the lessons we've seen here can prove useful. More generally, this idea of getting from one point to another in a constrained environment, like that of a maze, is important—the *network*, or *graph*, which is an alternative way of describing a maze where passages are edges and intersections are vertices, is at the heart of many of the applications that we use and rely on nowadays. An application that knows how roads are connected—OpenStreetMap, for example—can tell you how best to drive from your apartment to the beach; a Web site that knows how people, places, and things are connected—say, Google's Knowledge Graph—can produce better search results; a Web site that knows who your friends are—Facebook or LinkedIn, for example—can better guess who else you're likely to know; and a software system that knows how its components and modules are connected—say, Firefox—can better predict where future defects are likely to arise based on the pattern and density of connections.

Even robot vacuums serve as a good example. Not all robot vacuums are created equal, it turns out, because some are not as sophisticated as others when it comes to how much of a room they are able to cover. The simpler ones might roam around in random lines or in circles, whereas the better ones might map a room first, determining where the walls and any nooks and corners might be, and then move back and forth from one end to the other in a grid-like pattern. In other words, when a robot knows how best to get from one part of the room to the other, it maximizes its objective, resulting in a cleaner room.*

---

* In the March 2016 issue of *Cook's Illustrated*, there is a beautiful comparison of these different approaches, done by taking long-exposure shots of different robot vacuums in action.

BAD CHOICES

In the case of Ioannis, he will likely make it out this time with his sanity intact, irrespective of which method he opts for. However, if his hoarding continues to go unchecked and if his store keeps growing, he will have to start walking around with a ball of string in his back pocket.

5.

SORT THAT POST

Charlie Magna is already behind on his postal rounds. It is the middle of July in Cape Town, and the temperature is north of 45° C (113° F). To make matters worse, being the absentminded mail carrier that he is, he just knocked over his box of internally sorted mail, mixing all the bunches of letters that were meant for the thirty-three individual addresses on his rounds. To make matters even worse, he has photosensitive skin and he forgot his cap and sunblock at home. He is on his knees, on the scorching gravel, picking up the mail and trying to put everything in order again as quickly as he can before his skin suffers.

Hint: Consider how problems can be broken into smaller problems.

Order helps us do things faster. Imagine if a newspaper didn't list local events by day or if the episodes for a TV show that you were planning to binge-watch weren't listed in sequence. How annoying would it be to spend time rummaging for the next episode when that time could be better spent witnessing an ill-fated meth dealer get dealt yet another blow by a cruel universe.

Let us describe how Charlie might go about addressing his immediate dilemma.

OBJECTIVE: PUT THE SCATTERED BUNCHES OF ENVELOPES BACK IN THE RIGHT ORDER.

METHOD 1: PUT ONE BUNCH OF ENVELOPES ON THE GROUND IN FRONT OF HIM. TAKE A SECOND BUNCH AND IF THE ADDRESS IS CLOSER, PLACE IT TO THE LEFT OF THAT FIRST BUNCH. AND SO ON, UNTIL THE CLOSEST ADDRESSES ARE TOWARD THE LEFT OF THE LINE AND THE FARTHEST ONES ARE TOWARD THE RIGHT.

METHOD 2: LINE UP THE BUNCHES OF ENVELOPES ON THE GROUND IN FRONT OF HIM. SPLIT THE LINE SO THAT HALF ARE TO EITHER SIDE. SPLIT EACH OF THOSE HALVES, AND SO ON. FOR EACH PAIR OF BUNCHES, PUT THE CLOSER ADDRESS ON THE LEFT AND THE FARTHER ONE ON THE RIGHT, THEN DO THAT FOR EACH PAIR OF PAIRS AND SO ON.

And here is how those two methods look as a graph.

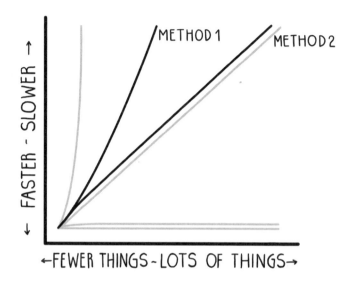

In real life, whenever we're sorting a number of things by hand, as Charlie is in this scenario, some variation of Method 1 is what we would likely use. As we have seen in the comparison graphs so far, this appears to be a general rule—for a few items, any method of undertaking the task at hand is probably fine. It is only when the number of items goes up that one method might become noticeably better than the other. Though Method 2 might not have a practical corollary in real life—in sorting at least—we'll discuss its general approach in conceptual terms.[*]

Notice first that Method 1 has a certain cadence to it. Charlie takes one bunch of envelopes, then looks through all the other bunches to de-

---

[*]   Not everything can be explained faithfully by analogy, nor should it.

termine where to put it. He then takes another bunch of envelopes and looks through all the other bunches, and so on. We have seen that type of approach before, haven't we? With the pile of socks. A notable difference is that for each envelope, Charlie looks at all the other envelopes just once, whereas with the pile of socks, Margie could potentially spend a long time searching for a match in the pile.

Charlie's approach in Method 1 is a signature *quadratic-time* algorithm.[*] Anytime you have a collection of things that you're searching through, be they of the same type or of different types, and you find yourself scanning that entire collection for each one of those things, you have a quadratic-time algorithm. Other examples of quadratic-time algorithms would be going through a pile of shirts and seeing which of the trousers in your wardrobe goes with which shirt, or looking through your grocery list and scanning the items on a store shelf to see if they have the item you need.

---

[*] Quadratic, since the time it takes one to sort the envelopes increases by an order of n^2 with the number of envelopes. In other words, an order of 100, say, seconds for 10 envelopes, 10,000 seconds for 100 envelopes, and so on.

SORT THAT POST

In computing, many of the simplest ways in which items are sorted run in quadratic time because, like Charlie's Method 1, they all happen to work by comparing adjacent items and potentially switching them around depending on which is bigger and which is smaller. We can in fact say with confidence that all approaches to sorting that follow this pattern of comparing adjacent items, run in quadratic time (n^2) on average. Put differently, if n is the number of envelopes, we might describe the function that puts those envelopes back in the right order by way of comparing adjacent envelopes as "bounded below by n^2," which is to say, on average (key word), we simply can't be faster than that. Some examples include *insertion sort, selection sort,* and *bubble sort.*

When I first learned about sorting as a sixteen-year-old schoolboy, I remember not quite understanding at first how one could do any better than quadratic time. The graph on page 45 suggests that Method 2 is noticeably faster than Method 1, so there does appear to be a way to sort in subquadratic time.

The general approach to a subquadratic way of sorting involves what's known as dividing and conquering, which is to say, breaking a collection of items into smaller collections of items and sorting those collections.* The breaking up of a collection into halves is logarithmic, as we saw earlier, and the putting of things back together again is linear, since we visit every item once. This approach to sorting is therefore said to be *linearithmic,* which you can think of as much faster than quadratic time and slightly slower than linear time.† Alternatively, it might be referred to as *log-linear* or simply *n log n*—the order we get from the time it takes to split the collection (*log n*) and to then put it back together again (*n*),

---

* The process implicitly involves a concept—called recursion—that we'll not really go into in this book but that you should certainly read about.

† Recall that logarithms grow slowly.

which when multiplied produce *n log n*. "Linearithmic" comes from combining the words "linear" and "logarithmic," resulting in a concept that suggests a degree of sophistication absent in the original parts. Much like Jedward.

Two well-known linearithmic algorithms are *mergesort*, invented by John von Neumann in 1945, and *quicksort*, invented by Tony Hoare in 1959. Charlie's Method 2 was akin to how mergesort works. The breaking-up step there involves separating the set of bunches of envelopes into individual bunches. And the putting-back-together step involves comparing and combining sets of bunches. After doing the latter step the first time, we would end up with sets of two in-order bunches. After doing it a second time, we would end up with sets of four in-order bunches, and so on. In Charlie's case, the process would look like this:

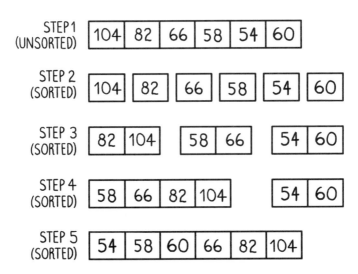

Notice how he goes from a set of unsorted envelopes in the first step to a set of sorted groups of envelopes, albeit of one size, in the second step. In each subsequent step, he further merges the envelopes creating larger sets of sorted envelopes until he is left with a single set containing all the envelopes. If we zoom into one of those steps, say step 4, we can see how the merging actually takes place.

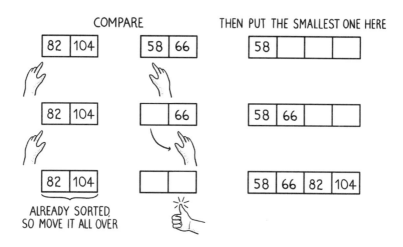

Method 2 is therefore the better choice given the speed improvement it affords. In the case of Charlie, his advantage is that he only has thirty-three bunches of envelopes to sort. Whichever method he goes for is likely to save him from a fortnight of inflamed skin. Had the bunches of envelopes been more numerous, however, Method 2's speed improvement would become more evident, and Charlie would undoubtedly benefit from knowing how best to sort his mail. For now, he is off to finish his round.

AS CHARLIE DRIVES OFF TO FINISH HIS ROUNDS HE COULDN'T BE HAPPIER TO HAVE LEARNED SOMETHING NEW TODAY. BEAUTIFUL MINDS ARE EVERYWHERE, HE TELLS HIMSELF. SOMETIMES, EVEN IN THE UNLIKELIEST OF PLACES.

"YOU'RE WELCOME."—TONY HOARE, INVENTOR OF QUICKSORT.

6.

BE MORE HIP

Foy is a recent transplant to the fine city of Ashland. Despite his carefully maintained goatee, and insistence on always appearing in public with a months-old copy of the *New Yorker* tucked under one arm—a copy he has never read, mind you, but which he likes to nonchalantly pick up and toss down in front of him whenever he's at a café or restaurant—Foy remains an outsider in his new home. His go-to response to any open-ended question, "Milton wouldn't be too happy, I can tell you that," is wearing thin. What do you think of Karl Ove Knausgaard's latest book, Foy? Milton wouldn't be too happy, I can tell you that. Did you enjoy Adele's new single, Foy? Milton wouldn't be too happy, I can tell you that. Alarmed by his crumbling facade, Foy has finally decided to overcome his faults and truly become one with culture and the arts. As a start, he would like to identify the world's most influential music and bathe in its glory but, overwhelmed by the choice in Ashland's record shops, he doesn't know what to do.

OBJECTIVE: LISTEN TO INFLUENTIAL MUSIC.*

METHOD 1: FIND AN INFLUENTIAL ARTIST AND LISTEN TO HIS OR HER SONGS, THEN FIND ANOTHER INFLUENTIAL ARTIST AND LISTEN TO HIS OR HER SONGS, AND SO ON.

METHOD 2: GO TO A MUSIC STORE, PICK A HANDFUL OF SONGS, AND LISTEN TO THEM.

---

* By "influential" we might mean music by artists who have inspired or given rise to other artists, or we might mean something much simpler, like popular. As we will see in a bit, what we're essentially interested in is answering the question, "Which of these songs is most important?"

Let us begin by characterizing how the soon-to-be-transformed Foy might go about his journey of gaining a new knowledge of music.

We might take Method 1 for granted nowadays because of the proliferation of recommendation engines on Web sites that rarely if ever present content to us unless it's somehow curated and tuned for our tastes. What Method 1 fundamentally comes down to is an approach known as *link analysis,* which says that if we have a collection of things that have some property in common, like songs or videos or people or components in a car, then by analyzing the relationships among those things—their links—we can answer such questions as "Which of these things is most important?" And that's precisely what we are interested in here. An example that many will likely be familiar with is citations. Often, a publication's citations, which is to say, the number of other publications that point to it, is considered a good indicator of that publication's importance. This approach of assigning higher value to the items that have the highest number of items pointing to them is the very approach that helped Google grow to prominence— it allowed its first-page search results to be much more relevant to users than those of the competition.

We will look at two types of relationships: the *degree* to which things are linked and how *similar* those linked things are to each other.

Degree: Say we have the collection of all the world's songs and we know or are able to determine for each pair of songs that one song's artist influenced the other song's artist. We might do that by looking at who has covered whom for instance or by appealing to some public data set. Our diagram might initially look something like the following:

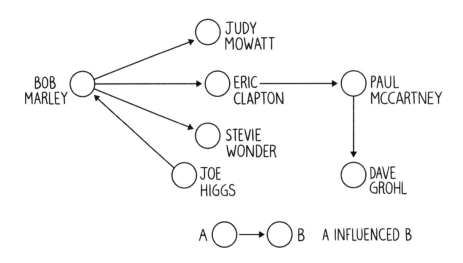

If we do that for all songs, we end up with lots of circles and lots of links, or a *network*. But that network is missing a crucial piece of information, namely, indirect links. If Bob Marley influenced Eric Clapton and Eric Clapton influenced Paul McCartney, then we'd like to say that Bob Marley's influence extends to Paul McCartney as well.

In order to capture those indirect links, we can perform a type of procedure known as *matrix multiplication,* in which we represent the artists in a square matrix, place a dot wherever an artist on the left has influenced an artist along the top and then raise that matrix to successive powers, each time capturing a deeper set of links. Once we can go no deeper, which is also known as achieving the matrix's transitive closure, we sum those matrices and get something like the diagram on page 57.

By counting the number of dots in each artist's row and looking at rows with lots of dots, that is, artists who have influenced a lot of other artists, we can identify a ranked set of highly influential artists and thus their songs. The procedure is domain agnostic—we could apply the ex-

	B	D	E	Jo	Ju	P	S
B		•	•		•	•	•
D							
E					•		
Jo	•						
Ju							
P		•					
S							

← BOB MARLEY HAS INFLUENCED THESE ARTISTS (5 OF 6)

↑

ARTISTS WHO HAVE INFLUENCED BOB MARLEY (1 OF 6)

act same steps to car parts, for example, by simply varying what we mean by a link. In the case of car parts, a dot might indicate a physical dependency, as in a wheel depends on an axle and an axle depends on the chassis. There, a practical application of link analysis would be to help answer a question like, "A lot of complaints have been coming in. We think it's due to a high level of complexity in this car model. Can you give me a list of the most highly connected parts?" The wheel is connected to two other parts, either directly or indirectly. Are there parts that are connected to four other parts, or five or ten? A greater number of connections in the case of artists is a good thing as it indicates influence. In the case of car parts, it could be a bad thing as it can signal a potential propensity for failures.

That can be useful as a starting point. At the cost of jumping between genres, we might choose to move from the most influential artist to the second most influential artist and so on. If at some point we're more interested in jumping from a song to another song that is similar to the one we are listening to, we will have to adopt a different approach.

Similarity: One of several ways to determine which songs might be similar to a particular song is to look at the artists behind those songs. And one way of determining which artists might be similar to, say, Bob Marley is to consider listeners—how many people who listen to Eric Clapton also listen to Marley? How many people who listen to Stevie Wonder also listen to Marley? And so on. If we do that for all artists and rank the resulting values from highest to lowest, we get a sense of which

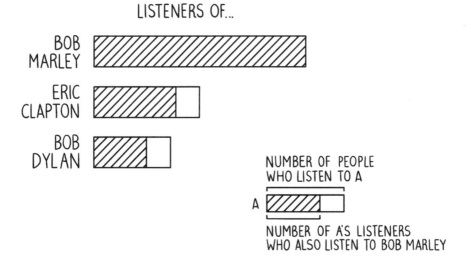

LISTENERS OF...

BOB MARLEY

ERIC CLAPTON

BOB DYLAN

NUMBER OF PEOPLE
WHO LISTEN TO A

A

NUMBER OF A'S LISTENERS
WHO ALSO LISTEN TO BOB MARLEY

other artists might be most similar to Bob Marley. The process is a bit more nuanced and can be improved upon using a number of more advanced methods. For instance, using what's known as a Jaccard index rather than simply counting, common listeners can help avoid results that might be skewed by artists who have many listeners.

You see the outcome of this form of analysis anytime you bring up the results page of a search engine or are presented with your news feed on a social media Web site, or are given recommendations about "items you might want to buy" on a shopping Web site or suggestions for "people you may want to connect to" on a professional networking Web site. Newspapers do it too, by modeling an article in terms of, say, the topics that it addresses, and then consider how similar that article is to other

ones. And video-streaming services' entire competitive advantage might rely on predicting what a subscriber likes and recommending content that is similar to that. In a recent blog post, Netflix fleshed out what signals they take into consideration when recommending movies and TV shows, which include not only the type of content—"You're watching a science fiction show, you might like this other science fiction show"—but also the region in which a person is located—"You're watching a cooking show, but you're in India, so you might like these Bollywood movies." An estimated 80 percent of Netflix's content views are the result of their recommendation engine. As we saw in chapter 4, connections of this sort coupled with the right analysis can afford insight.[*]

Method 2 ought to be self-evident, as it's essentially an uninformed pick. If we're at a record store, we might walk up to a box and pick out a handful of records. As with all uninformed picks, we have no way of knowing how much closer each pick gets us to our desired objective. Even if we were to come across music that happens to be influential, we would have no way of knowing that.

By working with the outcome of link analysis, we no longer have to make random guesses about where to start in the world of things we're interested in. If Foy were to take advantage of this advancement in technology, we might therefore characterize the two methods available to him as looking something like the following graph, with Method 2 taking a linear amount of time in the worst case and Method 1 taking a constant amount of time. Method 2 is linear, at worst, because Foy would

---

[*] It's important to consider the practical long-term effects of recommending similar things to someone. In the case of Foy, it makes sense, given his objective, but more generally, is it always a good thing to be watching the same types of shows? To be reading the same types of books, to be listening to the same types of pundits? Does that not preclude one from experiencing the full current of life? Algorithms are a reflection of the people behind them. One should always be cognizant of human biases that exist not only in what we say and do, but also in what we make.

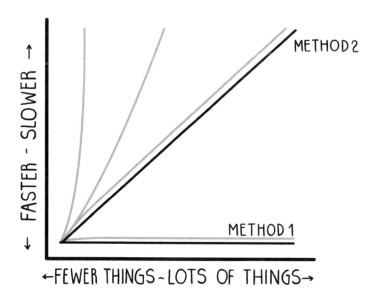

have to potentially visit all the songs in the world before he finds the ones he cares about, and Method 1 is constant because irrespective of the number of songs there happen to be in the world, Foy begins his journey within the circle of influential songs.

To appreciate how universal the application of Foy's task is, let's consider an example from a completely different domain—politics. Up until the nineteenth century, American politics looked quite different. At elections, streets were thronged with men (women weren't able to vote till 1920) who came out to parade, drink, and cast ballots. By then, voting had become less of a social event, which meant that politicians had to go out and look for voters. In the 1890s, William Jennings Bryan created what was perhaps the first example of a mailing list of supporters, a *database*, for use throughout his political career. In the twentieth century, those types of databases became more ubiquitous, and by the twenty-first century they were streamlined at the party level and gained the ability to better target people based on such nuanced characteristics as buying habits.

What this centuries-long development shows us is that in order for the political parties to canvass more effectively, and ultimately to save money and time, they had to know where to start with the electorate. Rather than blasting messages to the entire country, they found it more efficient to target people who were more likely to engage with them. That, as it happens, is a generalizable problem and the approach to tackling it is equally generalizable. So relevant is that approach, in fact, that it affects nearly every one of the major Web sites and services that you are likely to use today.

What does all of this mean for Foy? Is he now as cultured as he wishes to be? We have no way of knowing, but our goal has been to help him with the journey rather than the destination. One of the pitfalls of learning something new is starting at the wrong place, which can lead to setbacks like becoming disillusioned with the subject matter or giving up. The result of link analysis, an innovation boosted by the Web and perhaps soon by physical devices in the home and office that can talk to each other, is one way that curious people like Foy can acquire new qualities without becoming overwhelmed or going down an unnecessarily tortuous path. In Foy's case, his choice to abandon his old Luddite ways and to embrace technology that analyzes millions of songs for him will help him join the ranks of the cultured much sooner than he otherwise would have. He has already mustered the courage to sign up for a few local meet ups of music enthusiasts, so things are looking good.

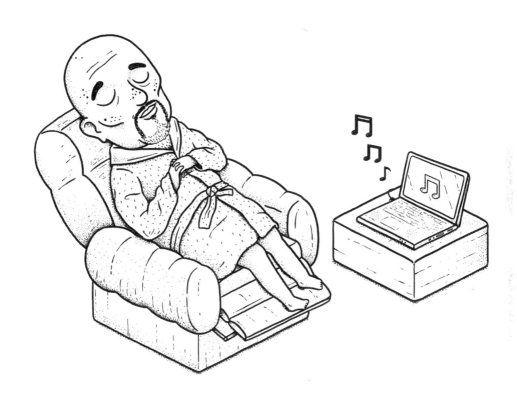

SOAK IT ALL IN, FOY. LIFE'S DESCENTS, AS THAT
HEAVILY MOUSTACHED GERMAN PHILOSOPHER MIGHT CALL
THEM, ARE SELDOM FACILE. BUT THEY ARE REWARDING.

7.

UPDATE THAT STATUS

Dwayne is on a walking tour of the Canadian Rockies. Crystal clear turquoise lakes lie motionless as their surfaces reflect with pinpoint accuracy the extraordinary grandeur of the mountains and trees above. Birds chirp and swoop beneath a clear sky while a gentle breeze makes its way toward the east. In that moment, the world seems like the most tranquil of places. No rivers of blood. No climate change. No poverty. No Kanye West. "The idealists were right," a fleeting cloud whispers from beyond the horizon. "They were right." And yet, Dwayne's mind is elsewhere. You see, a peculiar sight caught his eye that morning, just as his group was heading out of Vancouver. He witnessed a mallard wiggling around as though it were dancing the rhumba, and he has been struggling ever since. Struggling to find a witty yet succinct sentence to capture the hilarity of that scene that adheres to the 140-character limit imposed on him by the very tool meant to set him free. He cannot risk letting down the crowd of strangers whose approval he so relies on.

The brain, as they say in neuroscience, has the ability to detect salience. When you're in a quiet room and hear a noise of any kind, your brain will pick it up. If you're in a noisy room and hear a sound that is unlike those noises, your brain will pick it up too. In a sense, information that occurs more often is considered less meaningful, which is why the brain knows to filter such input.

The way in which some of us write text messages, by omitting frequently occurring letters like vowels, comes partly from an idea in information theory, where it is said that "only infrmatn esentil to understandn mst b tranmitd." Because of various properties of the English language, the previous sentence can be fully understood despite the omitted letters, which are predictable and therefore redundant and may therefore be left out. It is for that reason that when we're faced with the dilemma of shortening a piece of text while ensuring that it remains understandable, as Dwayne is, adopting such an approach isn't all that bad of an idea. In fact, we used to do it all the time before the advent of predictive text.[*] In return for saving space, this approach does result in lost data, albeit nonessential data. Note that all the topics we have talked about so far have involved faster and slower ways of doing something, whereas here, we're talking about things taking more and less space. This balance reflects the typical balance we see in practice when it comes to evaluating different approaches to solving problems—often, computer scientists compare the relative *speeds* of competing approaches, also known as their *run-time complexity*, but occasionally, they will compare them based on how much memory or disk space they require, also known as their *space complexity*.

---

[*] Nostalgic people in the comments sections of popular video-sharing Web sites have taken it upon themselves to carry that torch for posterity.

OBJECTIVE: REWRITE A WITTY STATUS UPDATE SO THAT IT'S 140 CHARACTERS OR FEWER.

METHOD 1: SWAP LONGER WORDS FOR SHORTER, BUT LESS CLEVER, WORDS.

METHOD 2: OMIT FREQUENT LETTERS, SUCH AS VOWELS, IN SOME WORDS.

What's fascinating is that Method 2 has an analog in computing. In 1952, computer scientist David A. Huffman came up with a method that reduced the amount of space required to store data. Unlike the earlier examples, Huffman's approach didn't involve removing things but rather focused on optimizing things as we will see in a bit. The way computers store data such as words is by mapping the letters in our alphabet—as well as numbers and other characters—to a set of numerical values. Those values are then stored using a representation that the computer can understand called *binary*. Each character is represented using a binary code that may consist of seven bits. So for instance, the letter "a" is mapped to the value 97, which looks like this in binary:

1100001

The letter "b" is mapped to the value 98, which looks like this in binary:

1100010

If we wanted to therefore represent the word "hans" in binary, it would look like this, with each letter taking up seven bits for a total of 28 bits:

1101000 1100001 1101110 1110011

The nice thing about characters having binary codes with the same length (seven bits in this case) is that it makes decoding a binary string easy. All we have to do is read off every seven bits and then use a table of mappings to decode it to English.

Huffman was a maverick, though. He looked at those seven bits and said, "Surely, there must be a way to compress things." His friends pleaded with him. "No, Huffman," they said. "It can't be done, Huffman.

It's too much to ask of one man, Huffman. Don't be a hero, Huffman." But Huffman didn't care. He was willing to propel himself into the unknown if it meant that he could potentially come up with an optimal binary representation for a set of characters.

Rather than using fixed-length binary codes, Huffman opted for variable-length ones. He exploited the fact that some characters in a sentence appear more often than other characters, and so he mapped the more frequently occurring letters to smaller values, and hence shorter binary codes, and less frequently occurring ones to longer binary codes. For example, say that we have determined that for a given body of data, the frequency distribution of characters is as shown on the left.

e	705
a	605
n	431
h	250
l	242
s	217
f	100
j	59

In other words, the letter "e" occurs 705 times, the letter "a" occurs 605 times, and so on. Notice that the characters are sorted from top to bottom in order of most to least frequent. Huffman's approach takes the pair of characters with the smallest frequencies, sums their values, storing the result in a new temporary character, and then sorts the set. It repeats the process until we have no more pairs of characters.

What we end up with is essentially a tree, where each node (a character) is connected to the pair of nodes that it came from with two edges. If we do that with the above set of characters, we end up with something like the following diagram, where we first pair "f" and "j," then we pair the result of that with "l," and so on. Each column, starting from the second one, is one step of the algorithm.

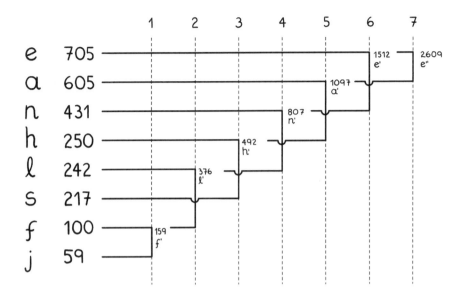

When we rearrange the diagram as a tree, it all becomes clear. A character's optimized binary code is the string that we get when we read off the bits from the topmost node—the *root*[*] node—down to that character's node. So in the tree below, any time we move left, we add a zero to a character's binary code and anytime we move right, we add a one to it, which is why the letter "e" ends up with the two-bit long binary code 11 and the letter "f" ends up with the five-bit long binary code 10001. Assigning a one or a zero to a node's children in the Huffman tree is done arbitrarily; that is "e" could have been coded as 01 instead of 11. While binary codes aren't guaranteed to be unique, they are guaranteed to be optimal. In any case, the Huffman tree is sent to the receiver along with the message so that the receiver knows how to decode the message.

---

[*]   In computing, trees are drawn with their root at the top and their branches extending downward, unlike actual trees.

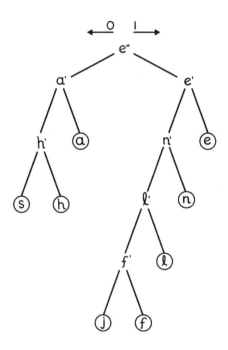

Here is a list of our optimized binary codes. Notice how the more frequent letters now have shorter codes:

e	a	n	h	s	l	f	j
11	01	101	001	000	1001	10001	10000

How does the word "hans" now look in binary?

001 01 101 000

We only needed 11 bits for that rather than 28 bits. This insight brings to mind an innovation from nearly a century earlier—that of how best to transmit messages between electric telegraphs. There, Samuel Morse's way of deciding which letters were represented using which codes was also based on how frequently letters occurred in the English language. Interestingly though, Morse determined the frequency of various letters not by talking to experts or conducting a study or analyzing data, but by counting the number of types in a printer's type box. So the next time a pedant questions your research methods, take heart.

Data compression techniques like Huffman coding are immensely important in the real world. Making optimal use of space means that Web sites load faster—Web servers can compress files before sending them across the network, and modern Web browsers can decompress them. When bandwidth is scarce, any such speed gain is critical. Compression also means that movies (think MPEG-2), pictures (think JPEG), and songs (think MP3) can all take up less space than they otherwise would have, saving money on their storage and transfer. Audio formats like MP3 are quite interesting in that their compression model relies on targeting audio components that humans can't hear due to biological or neurological limitations. For instance, the human ear isn't able to hear sound frequencies above 20,000 Hz.

The next time you have a voice or video conversation with someone without any drop in quality, think of the contribution that compression has made to that activity. Technology has matured to the point that your application only needs to send some data over the network and can then either infer or reconstruct the rest on the other end. In effect, compression helps lower the barrier to use of a technology.

That's all well and good. But what about Dwayne? Dwayne. Dwayne and his followers. Are they to be satiated? Luckily, yes. The boy has succeeded. He has successfully essentialized the scene. The event, its humor, and the important message it conveys for all who lend an ear. The power of the written word. This is what William Tyndale burned at the stake for. The boy updates his status and hundreds of thousands of eager users get the hit they were waiting for.

> Been on a fantastic hike since the quack of dawn. Bst part is som of the ducks have been putting on a show. Not looking forward to the bill!

One hopes that Dwayne's knowledge of how best to compress a message without any loss of essential qualities will always lead to similarly happy outcomes. If not, future mallards are bound to be close by.

8.

GET THE JOBS DONE

Kwee Noah works at the St. Louis office of agricultural biotech firm Hyperbowl, which specializes in selling genetically modified seeds. She's a lowly secretary and her boss has given her an impossibly large number of tasks to finish by the end of the week. The debrief is scheduled for Friday. That's two days away. If Kwee doesn't finish all her tasks on time, she won't be allowed to attend the end-of-year company banquet on Saturday. The banquet is the only event at Hyperbowl in which executives and staff rub shoulders. It is Kwee's best chance to see and be seen, and if she's lucky, it might help get her that promotion she has long hoped for. What will Kwee do?

Getting through a set of tasks is one of the most important responsibilities of an adult human. Think about how many books and articles you might have come across on how to get things done efficiently, how to avoid putting them off and how to become better at managing your workload. As the ever-diligent Kwee knows well, meeting a potentially unreasonable deadline can be stressful. Let's consider a few ways in which she might go about achieving her objective.

OBJECTIVE: FINISH ALL HER TASKS BY THE END OF THE WEEK.

METHOD 1: WORK ON A TASK FOR A BIT, THEN SWITCH TO ANOTHER ONE. WORK ON THAT FOR A BIT, THEN SWITCH, AND SO ON.

METHOD 2: SORT THE TASKS FROM EASIEST TO HARDEST. START WITH THE EASIEST ONE. ONCE THAT'S DONE, MOVE TO THE SECOND EASIEST ONE, AND SO ON.

METHOD 3: SORT THE TASKS BY PRIORITY. START WITH THE HIGHEST PRIORITY ONE. ONCE THAT'S DONE, MOVE TO THE SECOND HIGHEST PRIORITY ONE, AND SO ON.

All these methods ought to be familiar. With Method 1, we use chunks of time as our trigger for determining when to switch from one task to another. Say for instance that we have three problem sets for three different classes that we need to submit by the weekend. We might spend the morning on one, the afternoon on the second, and the evening on the third. We would then go back to the first one the next morning, and so on until we're done. This method of time slicing is in fact how modern operating systems handle multiple applications, and it is sometimes known as *context switching*. A scheduler looks at what processes[*] are currently running, it assigns them time slices, and then makes sure that each process runs for its allotted time. So seamless is the switching between processes that the operating system gives the illusion that processes are running in parallel. Nowadays, with multi-core processors, processes do indeed run in parallel. A processor with four cores might run four processes in parallel and need to context switch only when it has more processes than available cores. This brings to mind an interesting application of parallel processing that has an analog in real life, known as *pipelining*, in which a set of interrelated tasks are broken down and then carried out in a way that might optimize available resources. Say you and two friends just remembered that no one prepared the goody bags for the party that's about to end. In order to maximize the number of prepared goody bags per unit of time, it can be more efficient to adopt a sort of assembly-line approach—wherein you might write a greeting of some sort on a bag, your first friend might put the gifts in and a second friend might tie a ribbon around the bag—versus alternative approaches that might result in one or more friends idling while you finish writing on all the bags. Breaking tasks down can be an important

---

[*] Each application that you run spawns one or more processes that your operating system then manages.

part of efficiency, up to a point, of course—nine women cannot make a baby in one month, as Fred Brooks once wrote.

Given the current capabilities of hardware, we don't really feel the overhead of this context switching. In reality though, every time an operating system context switches, it has to hang on to the last process's state, clear up registers and transient data, and then load in the new process's state.* With humans, the cognitive overhead that this type of switching entails can be quite significant. Time and time again, we find that one of the biggest hindrances to productivity is having to stop what we're doing, perhaps in response to an urgent request, work on something else, and then come back to what we were originally doing. As it happens, operating systems do that too, by way of what's called an *interrupt handler*, which can temporarily stop a running process so that it can, say, either hand the process some data that it was waiting for, or to deprioritize it.

If the interruption is long enough, we might struggle for a bit as we attempt to get into the right state of mind for that original task. On the plus side, context switching does have the advantage of ensuring that we at least work on some part of each task. All the same, when it comes to humans in a time crunch, a "No Task Left Behind" policy can prove to be an idealistic one, leading as it does to widespread disappointments and a potentially disenfranchised populace.

Method 2 is an approach that procrastinators ought to be well acquainted with. It puts off the hardest tasks until the very end, opting instead to tackle simpler tasks first. It's an opportunistic approach that has the advantage of getting the greatest number of quick wins, potentially at the cost of bigger ones later on. The approach is sometimes re-

---

* Note that in most modern browsers, each tab also runs as a separate process, which is why you might see multiple entries for the same browser on your Mac's Activity Monitor or your PC's Task Manager.

ferred to as a *greedy* approach, a term that isn't necessarily meant to be disparaging, but to highlight the fact that the approach tries to get as far as it can with the least amount of effort.

One application of greedy approaches is trying to get from one point to another, say from one town to another town, in the shortest time possible. With a greedy approach, we would ask ourselves in each town, "What is the closest town from here?" It's quick-and-easy decision making, albeit it might prevent us from considering overall shorter paths, since the approach is biased toward shorter paths early on in a journey. We saw a version of that in chapter 4. Although with Ioannis, he just cared about getting out, not how long it took him. A lot has been written on shortest-paths algorithms and several competing ones exist. One of the more well-known approaches is called *Dijkstra's algorithm*. It was published in 1959 by Dutch computer scientist Edsger W. Dijkstra. More generally, this class of algorithms is known as *graph-search* algorithms.

Yet another application of greedy approaches that's worth pointing out is pattern matching, where a series of characters—a *pattern*—is used to search a body of text for potential matches, often in an attempt to replace those matches with something else. Say a body of text includes the line "Jessa Jessica" and we're interested in consolidating all the different spellings of that name. Given a greedy matcher, the pattern "Look for 'Jess' followed by any number of other letters followed by the letter 'a'" would return that entire line segment ("Jessa Jessica") rather than the individual names ("Jessa," "Jessica"), since the matcher would stop at the last "a" rather than the first one. This is sometimes desirable, though in this case it isn't so.

It's not too difficult to imagine actually doing something like this on an unfamiliar road—you're driving along an interstate and a sign tells you how many miles it is to the three closest towns. You might be tempted to head toward that closest town.

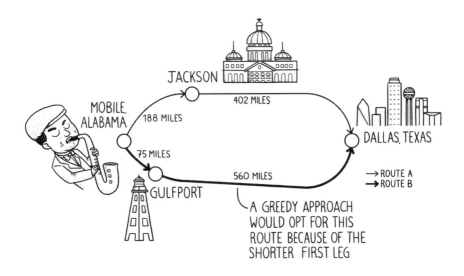

JACKSON

402 MILES

MOBILE, ALABAMA

188 MILES

DALLAS, TEXAS

75 MILES

560 MILES

→ ROUTE A
→ ROUTE B

GULFPORT

A GREEDY APPROACH WOULD OPT FOR THIS ROUTE BECAUSE OF THE SHORTER FIRST LEG

A non-greedy approach is undoubtedly more sophisticated and often leads to better results. We see an interesting analog for such an approach in military maneuvers where an immediate win, like securing one's own capital city, is abandoned in favor of a greater win later on—think Russia's handling of Napoleon's Grande Armée in 1812. A non-greedy approach might therefore be described as playing the long game. The *Wall Street Journal* published an article recently about the rise of Scrabble champions in Nigeria, and how their winning edge isn't necessarily due to a better vocabulary but rather to the counterintuitive tactic of opting for shorter words. Rather than playing seven- and eight-letter words, which are worth more points, the Nigerian Scrabble players found that playing four- and five-letter words resulted in a better overall strategy, seeing as the players were able to keep their most useful letters for upcoming rounds and face fewer potentially bad draws from the bag. This excerpt from the article paints a striking image of the beauty of such an approach:

> The Brit broke into a lead with AVOUCHED—an eight-letter bingo for 86 points—but spent the next five rounds managing awkward racks, playing words that scored in the low 30s and high 20s. With QUIZ (93), Mr. Jighere popped ahead. At the final score, which was 449 to 432, the winner's teammates lifted their champion around the room, singing a Nigerian pop tune: "We Done Win."

Method 3 mitigates one of the drawbacks of Method 2 by focusing on tasks that are actually critical rather than tasks that are perhaps lower impact because they are less involved. We place tasks or events or whatever else we might have into a list and order them by priority. Note that "priority" might in fact be a function of "time to completion," which is why it can be said that Method 2 orders tasks by priority too. An application where this equivalence makes sense is your printer. If a printer has a queue of ten fifty-page documents followed by a one-page document, it might make more sense for the printer to prioritize that last job rather than have it wait until the very end. Separating the two methods here is to make the point that a priority can rely on properties other than time.

If a new task were to come in, rather than adding it to the end of the list, we might determine that it goes somewhere in the middle because of its priority. You can imagine how adding tasks in the middle of an already-prioritized list might end up taking an increasing amount of time as you erase other tasks to make room for new ones. In chapter 12, we will talk about how a computer might choose to store such a list, which is often referred to as a *priority queue,* so that inserts can happen fairly quickly. Often in life, this approach is the most fruitful. Here is how those three methods look as a graph:

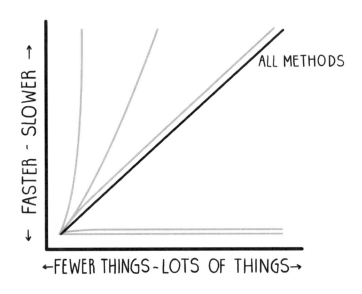

Assuming that tasks are independent of each other, which is to say the order in which we undertake earlier tasks is inconsequential to the time it takes to complete later tasks, all three of Kwee's methods take the same length of time to complete.[*] As mentioned in chapter 1, we are comparing the fundamental operation in all three methods, which is how long Kwee spends working on the tasks. Had we been considering instead, say, the time it takes Kwee to construct and maintain her collection of tasks, then we might have said that Method 1 takes a constant amount

---

[*] This may not always be true.

of time, namely zero, and Method 2 and Method 3 take a logarithmic amount of time at worst. Why focus on one set of choices over the other? Probably because given the costs involved in helping Kwee achieve her objective, we have figured that the first set of choices perhaps contributes the most to that overall cost compared to the second, which is to say, we have deemed that in practice, constructing and maintaining a collection of tasks isn't a big deal.* More on that in chapters 10 and 12.

This scene is a great example of when looking at just relative magnitudes is insufficient and when considering the actual outcome of a task becomes essential. The same applies to constant-time approaches. Imagine being a valet parking attendant and wanting to cram as many cars as you can into your parking lot so as to make the most optimal use of space. All your approaches for retrieving a customer's car might be constant time, and yet, the level of service that customers experience might vary significantly. For instance, not leaving a free lane through the lot to the exit might mean that the attendant has to remove eight other cars in the worst case to retrieve a customer's car. Leaving a lane free at all times might mean removing just three cars in the worst case. And imposing restrictions on when cars can be dropped off or picked up through a policy such as a "no in-out privilege," one might result in not having to remove any other cars per retrieval.

Failing to meet her overall objective, what is the next best outcome for Kwee? If it so happens that she cares most about high-priority tasks, what if the first task that she takes on is so unwieldy that she gets stuck on it all week? Might we perhaps combine a priority-queue

---

* This too may or may not be true. Consider the task of gathering all the tennis balls on a court after a practice match. Is the total distance you walk more important to you or the number of times you bend down? For some, it might be the former, in which case they might deem the shortest path that passes by all the balls to be the fundamental task. For someone else, someone with a bad back perhaps, it might be the latter, in which case he or she might opt for an alternative approach such as hitting the balls toward the net and then picking them up all at once.

approach with a context-switching one?* In the face of either ambiguity or idiosyncrasy, questions like these can lead to solutions that are creative and original. As an aside, in Keigo Higashino's 2005 novel *The Devotion of Suspect X*, a math teacher talks of making a geometry problem on a test look like an algebra problem to see how susceptible his students are to their own blind spots. It's a fascinating reminder of how easy it is to take things as read and to not challenge underlying assumptions. We tend to interpret new information in a way that fits with what we already believe. It's what computer scientist Alan Kay calls relativizing. This tendency can, of course, be either a vice or a virtue depending on how it is put to use.

---

* Douglas Carl Engelbart, inventor of the computer mouse, has written about the need to not only make the process of completing a set of tasks more efficient, but also the process of questioning whether the tasks are essential to begin with.

9.

FIX THAT NECKLACE

Jo is an independent crafter at a New Mexico arts and crafts market—Indian markets, as they are often advertised along Interstate 40. She has been suffering from rheumatoid arthritis for a few years now, making it ever more difficult for her to earn a living. The cruel hand of fate has resigned her to a job that relies on her being able to work with her fingers, selling personalized letter-bead necklaces. Jo's stall is situated right next to the entrance of that particular market, and as such, she makes every effort to convince any visitors who walk in that the best gift they could possibly buy a loved one back home is a personalized necklace. A little girl is compelled. She walks up to Jo. "Jacqueline, please." Jo gets to work, stringing beads onto an unassuming piece of twine and then gluing a clasp to both ends. She hands the little girl her necklace, but the little girl makes a face. She is unhappy. "Sorry, but I spell my name with an 'X' after the 'Q.' The 'X' is silent. All the trendy kids have unique names these days, don't you know?" Poor Jo.

BAD CHOICES

In chapter 1, we talked about arrays as a way to store a collection of things that can be scanned in linear time. And in chapter 3, we saw that it was possible to have approaches to tasks that take a constant amount of time irrespective of how large the task is. In this chapter, we'll talk about an approach that is focused not on scanning a collection of things but on allowing us to add and remove things to and from a collection at arbitrary points and in constant time. First, let us consider two ways in which Jo might fix the little girl's necklace.

OBJECTIVE: ADD THE MISSING LETTER BEAD TO THE NECKLACE.

METHOD 1: UNCLASP THE NECKLACE, REMOVE THE BEADS ONE BY ONE UNTIL SHE GETS TO THE "Q" OR "U," ADD THE MISSING "X," THEN ADD BACK THE OTHER BEADS ONE BY ONE.

METHOD 2: CUT THE NECKLACE BETWEEN THE "Q" AND "U" BEADS, INSERT "X" INTO EITHER END OF THE STRING, THEN TIE THE STRING WITH FABRIC GLUE.

One of the downsides of arrays is that they store things in contiguous blocks, which is to say, things that appear next to each other are quite literally stored next to each other in memory. And so, if it happens that we ever need to add something in between two things in an array, we can't just do that willy-nilly—we have to move everything after that spot over by one first so that we make space for the new thing and then we add the new thing. That's what Method 1 involves. Jo has to remove the beads one by one, from either end in this case, until she gets to the spot where the new bead is supposed to go. She then inserts the new bead and replaces all the beads that she took out. You can imagine how this process might take twice the amount of time for a name that is twice as long.[*]

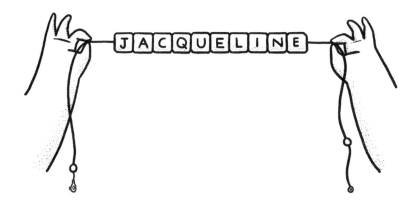

---

[*] Twice as long in the worst case, which occurs when the missing letter is somewhere near the middle.

Method 2's innovation is that it exploits the fact that Jo is holding together her beads with a piece of string, which can be cut at any point and then fixed by tying the loose ends in a knot or using fabric glue. This is an important property of the string because, as we'll see in a bit, it allows us to completely mitigate the issue that arrays have, where adding or removing things involves paying a high cost. Up until this point, notice how Method 1 might come across as the more attractive method—what's removing three or four beads versus having to cut the string and then tie its two ends? Consider whether this attractiveness would still hold had we a larger number of beads.

In computing, there happens to exist a structure that surfaces precisely this property, and this is how it looks:

$$J \rightarrow A \rightarrow C \rightarrow Q \rightarrow U \rightarrow E \rightarrow L \rightarrow I \rightarrow N \rightarrow E$$

Notice that we still have a collection of things, but now we're no longer constrained by the need to store things contiguously. Rather, each thing in our collection is simply pointing to the thing after it. This reference—or *link*—between each pair of things is analogous to Jo's string. And so now, if we want to add an item somewhere near the start of a collection, we no longer have to worry about making space for it. We can simply modify those links. The same is true for removing an item.

$$J \rightarrow A \rightarrow C \rightarrow Q \rightarrow U \dashrightarrow E \rightarrow L \rightarrow I \rightarrow N \rightarrow E$$

$$\searrow X \nearrow$$

A → B   A LINKED TO B

A --→ B   REMOVED LINK
BETWEEN A AND B

This structure, first developed in the mid-1950s, is known as a *linked list* and is fundamental to many applications in computing because of its efficiency in dealing with inserts and deletes at arbitrary points in a collection. For instance, we mentioned in chapter 8 that a printer might use a queue to store its jobs, and it might decide to perhaps sneak in smaller jobs in between larger ones. An efficient way for it to do that would be to implement its queue using a linked list. As before, keep in mind that we are focusing on the fundamental operation here, which is the cost of adding or deleting an item in a collection. There are other operations at play here as well, such as looking for the item that we want to add the new item after. In most cases, it would likely take the same length of time to look for that item in both an array and a linked list.

Another application is text editors, which might choose to represent a document of text by storing the lines in a linked list so that when you move lines around or add lines in between other lines, all the text editor has to do is modify those lines' links rather than physically move the lines around in memory. A detail that we have glossed over here is that our links only go one away, from each item to the next. In some cases, like with the text editor example, a line would in fact benefit not only from knowing which line comes after it but also which line comes before it. That way, if our cursor is on a particular line, and we move the cursor to the previous line, our text editor can simply follow the link to the previous line rather than going back to the beginning of the linked list, known as its *head* by the way, and moving through it node by node. This modification to a linked list results in a structure known as a *doubly linked list*. The name seems to have been thought up by the same jovial person who named the walkie-talkie.

Jo is by no means a quitter. Even so, she will undoubtedly benefit from knowing the quickest way to fix a necklace after the fact. The way she might add the missing bead brings to mind how screenwriters modified their screenplays before the advent of word processing software, by cutting their typewritten sheets of paper at the right places and taping the new content to those sheets. Here is how Jo's two methods look like as a graph:

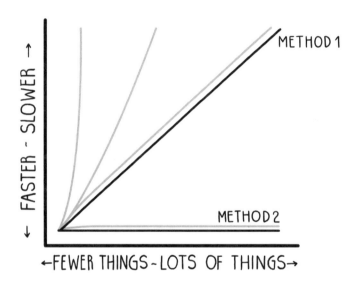

10.
LOCATE THAT BOX

Ludwik sells computer appliances out of his store on Mission Street. He lives close by on the fourteenth floor of a forty-two-story apartment complex, every shared space of which is surveilled using closed-circuit cameras. In order to make the requisite amount of money every month to offset his rising rent, Ludwik often picks up cardboard boxes from the recycling rooms in his building and uses them to ship memory modules to customers abroad. There is a recycling room on each floor of his building. He has one order that he needs to fulfill today and the post office closes in fifteen minutes. Ludwik needs to find a cardboard box for his package fairly soon.

OBJECTIVE: VISIT AS FEW FLOORS AS POSSIBLE TO FIND AN EMPTY BOX.

METHOD 1: GO FLOOR BY FLOOR LOOKING FOR AN EMPTY BOX.

METHOD 2: ASK THE FRONT-DESK STAFF IF THEY CAN INSPECT THE RECYCLING ROOMS' LIVE FEEDS.

Let us characterize how Ludwik might achieve his objective of finding an empty box in his building:

Method 1 is Ludwik's visceral approach—he starts at the very top floor perhaps and makes his way down, one flight of stairs at a time. The time it takes to undertake Method 1 could be cut in half by having a friend check even floors while Ludwik checks odd floors, albeit the approach might still be characterized as taking linear time for reasons we'll get into in a bit. Method 2 proves to be the better approach, seeing as it allows Ludwik to determine which recycling rooms have empty boxes in them by having the front-desk staff glance at a series of live feeds on a screen. In light of his objective, this approach allows him to potentially find an empty box in constant time rather than in linear time, since he

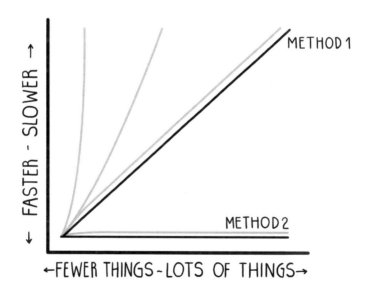

has to visit no more than the one floor. The call to the front desk is therefore a constant cost that Ludwick pays in order to avoid linear growth.

This might be a good point to talk about the way in which we are measuring rates of growth. In this book, we are purposefully abandoning some rigor for the sake of insight. Even so, it's important to realize that there are different ways in which we might describe the rate of growth of a particular approach, or *function*. One way is known as *big-theta notation*, which characterizes a function in terms of a set of upper and lower bounds—for a large number of things, it says that our function can grow *no faster than* some simpler function, like a linear function (*n*) or a logarithmic function (*log n*), and *no slower than* some other function.[*] Our function is therefore tightly bound by these other functions,

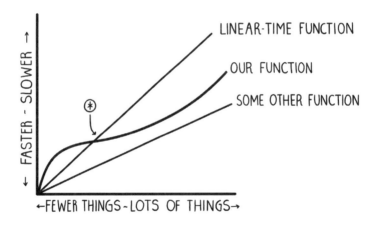

⊛ BEYOND THIS POINT, OUR FUNCTION IS ALWAYS BOUNDED ABOVE BY THE LINEAR-TIME FUNCTION AND BOUNDED BELOW BY SOME OTHER FUNCTION.

---

* Recall that the slower a function grows, the more attractive it is.

and that's the reason why we can say things like "binary search is better than linear search because it takes logarithmic time in the worst case." As we saw in chapter 2, binary search (a logarithmic time method) allows us to find a shirt among a rack of one hundred shirts in under seven steps and among a hypothetical rack of one thousand shirts in only ten or so steps. Contrast this with one hundred steps and one thousand steps, respectively, in the case of linear search.

There are two things that big-theta notation assumes. First, it omits coefficients, reasoning that their values become inconsequential as the number of things increases.[*] And so, with Ludwik, a rate of growth of either n or $n / 2$ would both be characterized as taking linear time and written $\theta(n)$, read "big-theta of n." Second, big-theta only considers the dominant term in a function because it assumes that the dominant term has a greater impact on a function's output relative to other terms. We have been calling that dominant term the fundamental operation so far. An example from computer science professor Mark Weiss elucidates this point:

> In the function $10N^3 + N^2 + 40N + 80$, for $N = 1,000$, the value of the function is 10,001,040,080, of which 10,000,000,000 is due to the $10N^3$ term.

And so, if Ludwik's Method 1 involved him visiting, say, his own floor twice, we would characterize the time it takes him to achieve his objective in the worst case as $t(n) = n + 1$, where "+ 1" indicates that extra visit, and write it in big-theta notation as $\theta(n)$. It is imperative to mention that this assumption comes with a series of warnings. For instance, there

---

[*] This, of course, may not be the case. There is always the possibility of large constants whose effects on a function may be nonnegligible.

might be cases in which nondominant terms have a noticeable impact on a function. Consider the case of Jo and her string of beads in chapter 9. We focused on the adding of beads and deemed her first method a linear-time one and her second method a constant time, and therefore, a more attractive one. In doing so, we took it as read that the gluing of the string's loose ends was a trivial task, but what if the glue required five minutes to set? Would that perhaps influence her choice? Such constant values are most noticeable when we are dealing with a small number of things, and they demand that we, at the very least, be aware of them.

Two other notations that complement big-theta and operate under the same assumptions are *big-omega*, which places a lower bound on a function for a large enough value of *n*; that is, it says that our function can grow no slower than that lower bound, and *big-o*, which places an upper bound on a function for a large enough value of *n*; that is, it says that our function can grow no faster than that upper bound. Of course, in reality, the function could grow slower than that upper bound and thus be more attractive, but that's big-o. Ever the pessimist. Ever the embodiment of Sod's law. In this book, when we talk about an approach's rate of growth, we are generally referring to its performance in the worst case when it is tightly bound, which is to say, its big-theta estimate. Note that any level of indirection like the one we have adopted in this book introduces a trade-off. And though we are reaping the benefits of that trade-off, it is important to know that reality is more nuanced. There is more on this topic to be found in the references listed at the end of this book. As far as Ludwick is concerned, the real-life difference between the two approaches is quite noticeable, making his decision an easy one.

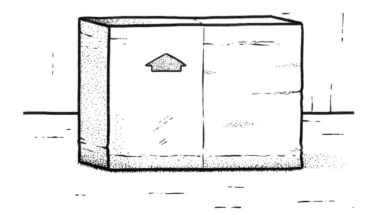

11.

FILL THOSE SHELVES

Terry is a sophomore at Medlock High, an exclusive school in Beverly Hills, California. He is currently serving detention for giving a needlessly provocative talk—titled "Not Everything Has to Have Avocado and Kale in It"*—in his social studies class. As punishment, his principal has instructed him to spend however long it takes at the school library, stacking books on a newly bought shelf after the old one buckled and sent close to 250 books crashing to the ground. All Terry has to do is put the books onto the shelf in alphabetical order by author's last name. Terry was hoping to catch a movie that evening with some friends, not be stuck at school stacking books. He might still make it, but he mustn't waste any time.

---

* The gall of some people is beyond measure.

OBJECTIVE: PUT THE BOOKS BACK ON THE SHELF IN ALPHABETICAL ORDER.

METHOD 1: PICK UP A BOOK AND PLACE IT ON THE SHELF. THEN PICK UP ANOTHER ONE AND PLACE IT ON THE SHELF, BEFORE OR AFTER THAT FIRST BOOK DEPENDING ON WHERE IT SHOULD GO, AND SO ON.

METHOD 2: USE BOOKENDS TO CREATE SPACES FOR EACH LETTER OF THE ALPHABET, THEN PLACE EACH BOOK IN THE APPROPRIATE SPACE, ADJUSTING THE BOOKENDS AS NECESSARY.

FILL THOSE SHELVES

Let's begin by characterizing how Terry might go about his task.

We saw in chapter 5 how approaches to sorting that rely on comparing adjacent items and deciding which is bigger and which is smaller take quadratic time. And we said that examples of that approach in practice include insertion sort, selection sort, and bubble sort, all of which operate using that blueprint with minor differences. Terry's Method 1 is in fact the exact same approach as Charlie's Method 1 from chapter 5.

What's interesting here is that Terry's improved method does not opt for a divide-and-conquer approach, as Charlie's did; it merely improves on his original quadratic-time approach. And it does so with a fairly simple innovation. If we take an approach such as insertion sort, we find that the greatest contributor to its slowness is inserting an item in the right place because every time we do that, we then have to shift all subsequent items over by one. What Terry's Method 2 does is that it preemptively accounts for those shifts by placing evenly spaced gaps at various points along the shelf. That way, when he does find the right spot for an incoming book, he will likely only have to move a few books over before he hits a gap. Terry is essentially trading space for time, and the bigger or more numerous his gaps, the fewer books he has to move every time. This improvement to insertion sort turns it from a quadratic-time approach to a linearithmic one, "with high probability" as its creator puts it, and is referred to as *library sort* or *gapped-insertion sort*. Recall from chapter 10 the idea that while the initial setup cost of placing the bookends on the shelf can make this approach slower when only a few books are involved, with a large enough number of books, it does eventually outdo the alternative approach.

FILL THOSE SHELVES

Here is how the two approaches look as a graph:

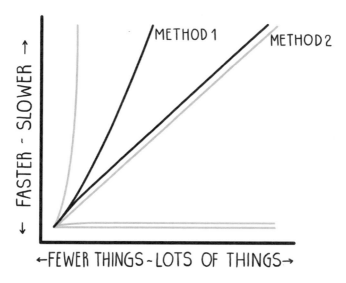

We can perhaps better see the value of such an approach if we consider an alternative scenario that is a bit more constrained. Consider the following grim picture that the authors of library sort paint—the tech industry has fallen, a fatal spear thrust into its liver by a world that is sick and tired of their silly ideas. As a result, thousands upon thousands of destitute tech workers have sought refuge in the one place on earth where the talents of those broken men and women are put to use in exchange for free pizza and condiments. The influx of talent has made universities across the country ecstatic. Everyone is thrilled except for the mailroom person at this one college who happened to graduate from Medlock High. Yes, it's that same Terry, fifteen years later. He manages a wall of cubbyholes that are labeled with the names of all the graduate students in that department in alphabetical order. Every time a new student joins, he has to add that person's name on the wall, possibly by replacing an occupied cubbyhole, and then move over all the other labels by one.

ADAMS	BARNES	BUTLER	COOK	DIAZ	HARRIS		
ALLEN	BELL	CAMPBELL	COOPER	EVANS	LEE		
ANDERSON	BENETT	CARTER	COX	FISHER	ORTIZ		
BALLEY	BROOKS	CLARK	CRUZ	GARCIA			
BAKER	BROWN	COLLINS	DAVIS	HALL			

Now that the frequency of moving cubbyholes around has sky-rocketed, Terry is reminded of the approach he took as a schoolboy a decade and a half ago. He recognizes that a similar approach to the one he adopted back then can help alleviate his stress at the cost of a bit of extra space. And so he proceeds to add empty cubbyholes in between various occupied ones.[*]

---

[*] The first language I ever programmed in was BASIC. During a recent conversation with a friend, I was reminded of a memory I had all but forgotten—that line numbers in that programming language typically had "gaps" between them. Line two didn't follow line one, rather line

The mention of "high probability" earlier brings us to an important topic and the main takeaway for this chapter. We have hitherto mentioned in passing "worst cases" and "average cases" for different approaches. These qualifiers are essentially indicators that give us a sense of how long we can expect a particular approach to take to go through all of its input. How long an approach actually takes to complete can depend on a number of things. For instance, in code, we might have logical branches in place (e.g., if this happens, then proceed, otherwise skip), and depending on what path is followed during a particular run, run times can vary. A good example of how input can impact run time is quicksort, which we mentioned in chapter 5. Quicksort begins its divide-and-conquer endeavor by first selecting one of its items to serve as the *pivot*. It then uses that pivot to separate items into two groups—one that contains items that are smaller than the pivot and one that contains items that are larger than the pivot. By choosing one of the items at random and to serve as the pivot, quicksort can run in linearithmic time in the average case. By choosing a pivot that's too small or too large though, quicksort runs in quadratic time in the worst case. This dramatic shift in performance is because when the pivot does not do a good job of halving the collection of items at each step, we end up having to inspect nearly every item at each of those steps, which as we saw in earlier chapters, is a signature of a quadratic approach.

In the case of quicksort, the analysis leads us to the belief that the algorithm has a high probability of running in linearithmic time. We know what we need to do to achieve this near guarantee—ensure that the pivot is never the lowest or highest value—and so the average case is for all practical purposes a good enough reason to assume that quicksort will run in linearithmic time.

---

twenty might have followed line ten, and so on. That convention allowed a programmer to add new lines in between existing lines while avoiding the drudgery of having to renumber them.

Sometimes, a probabilistic guarantee is still not good enough. If it so happens that our approach impacts a mission-critical application, like controlling a space shuttle, or a life-threatening application, like regulating an infusion pump or a defibrillator, then the worst-case indicator of an approach is likely the one we will care most about. Such a classification of potential outcomes is useful even in other applications in everyday life.

For Terry, an average case of linearithmic time is more than rewarding despite the fact that his approach runs in quadratic time in the worst case. He has nothing to worry about for now. As long as he can find enough bookends, or fashion some himself, he should be able to make it to the theater just in time.

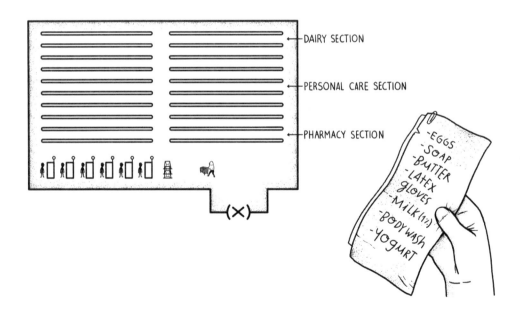

DAIRY SECTION

PERSONAL CARE SECTION

PHARMACY SECTION

(X)

-EGGS
-SOAP
-BUTTER
-LATEX GLOVES
-MILK (1½)
-BODYWASH
-YOGURT

12.

NAVIGATE THOSE AISLES

Wurzma Monet was until recently a hedge fund manager. His extraordinary decision to become a rapper was brought about right after he gave a talk about credit default swaps to a class of ten-year-olds. It was "What does my daddy do for a living day," and by noontime, Wurzma had made up his mind that he wanted to live a more fulfilling life. He visits the local supermarket every two weeks and often finds himself wandering between aisles from one end of the store to the other looking for items on his shopping list. Back and forth and back and forth. It takes him forever to finish shopping, a fact that is exacerbated by the awkward way he chooses to tread, which is less "gangsta walk" and more "I missed my appointment with the physiotherapist." Other rappers are beginning to notice and Wurzma's fragile "street cred" is slipping as a result. He needs to stop looking like such an amateur and being condemned to a lifetime of ridicule.

Wurzma is at an important juncture. The good news is that we probably already know how to help him, and so in this chapter we'll elaborate on some of the concepts that we talked about in earlier chapters. First, here are two ways that Wurzma might go about looking for the items on his shopping list at the supermarket.

OBJECTIVE: MINIMIZE THE NUMBER OF AISLES HE PASSES THROUGH.

METHOD 1: GO THROUGH THE SHOPPING LIST IN SEQUENCE.

METHOD 2: PREPARE THE SHOPPING LIST AHEAD OF TIME SO THAT IT IS ORDERED BY CATEGORY. GO THROUGH THE CATEGORIES ONE BY ONE WHILE SHOPPING.

YO, I put an array in your array so you won't go astray

So far, we have talked about arrays as a fundamental type of structure to store collections of things. In chapter 6, we introduced another useful structure called a matrix, which also stores a collection of things but unlike arrays stores those things along two dimensions rather than one. What the two structures have in common is that an array can in fact be turned into a matrix. All it has to do is instead of storing a *literal* (like a number or a letter or a word) at each location, it stores an array. An array of arrays is known as a two-dimensional array, or more generally a *multi-dimensional array*.

	0	1	2	3	4
0	LATEX GLOVES				
1	BODY WASH	SOAP			
2	EGGS	BUTTER	MILK (1%)	YOGURT	CHEESE
3	GARBANZO BEANS	OLIVE OIL			

What multidimensional arrays allow us to do in Wurzma's case is store his shopping list across two dimensions, thereby allowing him to iterate across each subarray based on the section of the supermarket he is in. And so, rather than his shopping list being a list of items, it's now a list of categories, and each category is in turn a list of items.* Since items in a store are typically grouped by category, Wurzma can just walk through to the part of the store that stocks, for example, "personal care" products, and then go through his "personal care" array and get those items from that aisle. And so on for the rest of the items. That's what Method 2 does. Had Wurzma just used a list—an array, as in Method 1—then he might have ended up wasting around thirteen minutes going from aisle to aisle seeing, as in the worst case, Wurzma passes by all the aisles for each item on the list. If n is the number of aisles and m is the number of items on the list, then $n / 2 \times m = (n \times m / 2)$. For twenty items and forty aisles, Wurzma passes by those twenty rows of aisles twenty times. Assuming it takes two seconds to pass by each row of aisles, wasted travel time going from aisle to aisle comes to around thirteen minutes. Whereas now his inter-aisle commute time is under a minute, since he visits no aisle more than once, seeing as, in the worst case, Wurzma passes by all the rows of aisles once at most; that is, $n / 2$. So for twenty items and forty aisles, Wurzma's wasted travel time going from aisle to aisle is under a minute.

Notice that while Method 1 doesn't exactly look like the quadratic-time sorting algorithms we saw in chapters 5 and 11, it does follow the same blueprint, since it causes Wurzma to potentially pass by all the aisles in the store for each item on his list in the worst case. Here is how the two methods compare:

* Any interesting data set that you might think of will likely be analogs to a multidimensional array, seeing as it would have multiple columns that a statistician would then use to perform various analyses on the rows (the observations).

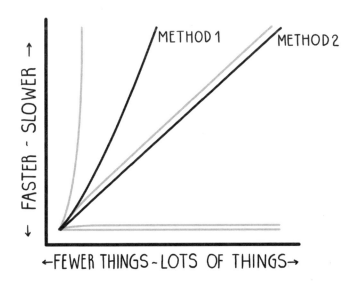

BuMP, buMP, buMP; do it often, you'll be sorry, like a character in Murakami.

In chapter 1, we talked about hash tables and how they are useful when we need to do rapid lookups and don't care about order. This talk of multidimensional arrays is a good reason to expand our understanding of hash tables. What we took as read in that earlier discussion was that a hash function always maps an item to some unique location in the hash table, and that is how it guarantees constant-time lookups. In reality, however, a hash function can encounter *collisions*, whereby it resolves to a location in a hash table that is already occupied by another item. This can happen either because the hash function isn't perfect, which is to say it doesn't do a good job of uniformly distributing hash values, or because we have more items that we want to store than the hash table has space for. The proportion of a hash table that is occupied is referred to as its *load factor* and ranges from zero when the hash table is empty to one when the hash table is full.

In such cases, one of the things that we might do to resolve the collision is known as *chaining*. With chaining, rather than have a hash table of items, we would have a hash table of collections of items. That way, when a collision occurs, the colliding item is pushed onto the end of the collection at that location and no data are inadvertently overwritten. And so, we end up with a hash table that is a collection of collections. Whenever our hash function happens to resolve to a location that has multiple items, we end up having to iterate over those items until we find the one that we're looking for. All this is of course completely transparent to the user.

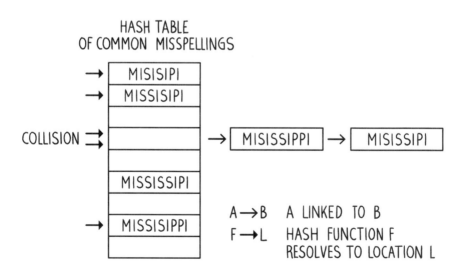

HASH TABLE
OF COMMON MISSPELLINGS

Wonderin' where to go next, don't worry, I'll tell ya where to head, so hurry.

Another thing worth pointing out is that the multidimensional array from earlier has, in effect, imposed a priority on our items; namely, the distance from the store's entrance to the particular aisle the item is in. Perhaps this would be a good time to expand our understanding of priority queues, which we mentioned in passing in chapter 8. There, we mentioned that when you jot down some sort of prioritized list of items and then have to add a new item to the list, you might have to erase some parts of the list to make room for it. Soon enough things get messy and you find yourself having to write out a new list. How might a machine manage such a list with the efficiency that we'd expect of it? We've already seen structures that are optimized for scanning (arrays) and inserting things at arbitrary points (linked lists). We'll now elaborate on the priority queue, which is optimized for surfacing the highest-priority item[*] in a collection in logarithmic time. It's called a queue even though it isn't what we might colloquially understand to be a queue, where the first thing to go into a queue is the first thing to subsequently come out.[†] Rather, you can think of a priority queue as a sort of underground plant that only releases one sprout at a time and makes that available to the passersby for the picking.

When we pick that top-priority task, the tree *rebalances* itself and pops out the second-highest-priority task, and so on. This particular

---

[*] Actually, it surfaces the minimum (or maximum) item, but we're using that property in this case to surface the highest-priority one.

[†] Having said that, a structure that satisfies such a first-in-first-out (FIFO) property does exist and even has analogs in everyday life. Consider the way grocery stores push perishable items onto a shelf, with older ones placed near the front and newer ones near the back.

way of describing a priority queue is referred to as a *heap*. We can't really explain a heap by analogy, but it remains a fascinating structure and is certainly worth our while to understand how it is able to rebalance itself and bring to the top the highest-priority item in logarithmic time and how it ensures that inserts can also happen in logarithmic time.

Let us first model Wurzma's prioritized list as a heap.

1. LATEX GLOVES
2. BODY WASH
3. SOAP
4. GARBANZO BEANS
5. OLIVE OIL
6. EGGS
7. BUTTER
8. YOGURT
9. CHEESE
10. MILK

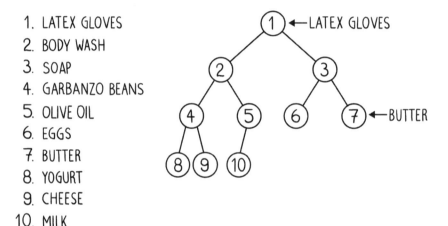

Notice that the heap is actually a tree of nodes and has two properties. The first property is that every node has a lower priority than its parent node.* That is why the highest-priority item, which is the closest item to the store's entrance, is at the very top. Nothing else is to be inferred about the order of other nodes, such as siblings, which is to say, nodes that are on the same level as each other. There are other structures

---

* Except for the topmost node—the root node—seeing as it has no parent.

that ensure that all the nodes in a treelike structure are ordered, such as a *binary search tree*, which can be useful in situations such as the one we described in chapter 2. The second property is that each node has two child nodes, with the possible exception of the lowest level nodes. This structural invariant ensures that the heap's height, which is to say the longest possible path, is never deeper than *log n*, where *n* is the number of items in the heap.

Here's what happens once Wurzma gets the box of latex gloves, which we've determined is the closest item to the entrance, and needs to know which item to pick up next.

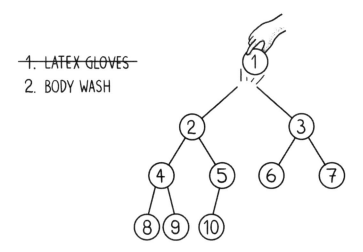

1. LATEX GLOVES

2. BODY WASH

Once we remove the top node, the heap's first property is broken, and so it invokes its rebalancing algorithm, which involves replacing the empty root node with the last node in the heap and then checking that new root node against the smallest of its children to see if it needs to be swapped. That checking and swapping is done between each

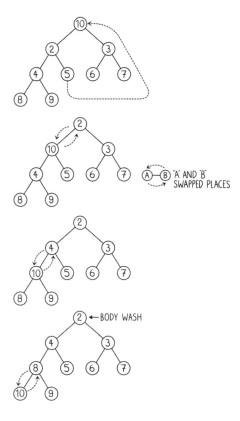

pair of parent-and-smallest-of-two-children nodes all the way to the last node in the heap. Notice that this process of rebalancing the heap takes logarithmic time[*] and results in the next closest item being at the top.

Similarly, if we were to insert a new prioritized item into the list, no erasing and moving of things about is necessary, beyond a *log n* number of moves. We add the new item to the end of the heap and then swap it with its parent node if the parent node happens to be larger than the new item. We continue swapping, if necessary, up to the root node.

---

[*] Recall that $n = 10$, and so $log_2 10 \approx 3$.

NAVIGATE THOSE AISLES

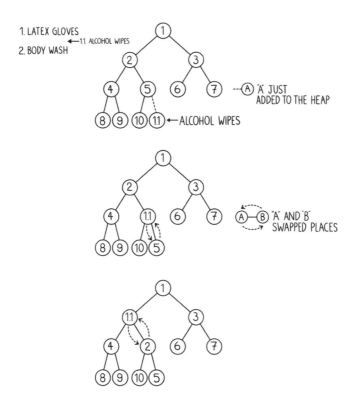

1. LATEX GLOVES
← 11. ALCOHOL WIPES
2. BODY WASH

A̅ JUST ADDED TO THE HEAP

← ALCOHOL WIPES

A̅ AND B̅ SWAPPED PLACES

We have spent these last few pages on elegant approaches that a computer might leverage when solving a problem. Mentioning them at the end of this book is to emphasize the fact that algorithms, much like equations in the sciences, can have all the properties that one might attribute to disciplines like art: beauty, elegance, and grace. If you happen to venture into more advanced realms where algorithms play an essential role, it is worth taking note not only of the outcome and performance of those algorithms but also of how they are modeled. Either way, venture into those realms and you might find yourself in the world of artificial intelligence, where real-world applications of algorithms include hospitals making quicker diagnoses and saving patients' lives, or researchers gaining a better understanding of complexities like the human genome. You might find yourself in the world of game theory, where real-world applications of algorithms include car-share companies making better decisions about how to match riders whose routes happen to overlap at a given time. You might find yourself in the world of computer vision, thinking about how to make self-driving cars good enough to become ubiquitous, or the world of image processing, thinking beyond features that rely on simple transformations, like brightness and contrast, to ones that rely on advanced algorithms that make various trade-offs, like how best to blur, sharpen, lens-correct, or color-correct images. The applications are truly vast.

As for Wurzma, his avant-garde rap career is about to flourish. Nobody has really thought of rapping about arrays, hash tables, and priority queues. How utterly cool is that? He has yet to tell his son, whom he plans to surprise by staging a rap battle at the poor boy's eleventh birthday party next week. What could possibly go wrong? More important, Wurzma has been getting much faster at shopping, so much so that his upcoming single is all boast. Good for him.

> *No more games, no jokes, I'm a new man, a phoenix*
> *Risen from the ashes, givin' it straight, like Bill Hicks*
> *I'm so fast you won't believe it, from the jam to the Kleenex*
> *Bread, milk and honey, organic sugar and then brownie mix*
> *I see you all staring at me, amazed and transfixed*
> *But you're too ashamed to admit it*
> *And that's ok*
> *You're the critics, you see*
> *And I'm Michael Bay's flicks*

FINAL THOUGHTS

When Feynman was once asked how he developed his legendary ability to solve problems so quickly, he said that it was because he modeled those problems in a multitude of ways, which is to say he looked at them from different perspectives. Analogies are one way of looking at a problem, equations are another, the dialectical method is yet another. For Feynman, ever the showman, a fantastically fictional story that was sure to astonish his audience was yet another.

There is an analog to that approach in art, where the ability to paint a scene from different perspectives without any loss of detail is considered a virtue. Interestingly, it was an innovation from Brunelleschi during the Renaissance that made that possible. By devising a mathematical way of codifying linear perspective, Brunelleschi was able to render a perfectly accurate drawing of the Baptistery of Saint John in Florence. Objects that were closer looked bigger, those far away looked smaller, and the lines of the baptistery's tiles converged in a way that was in accordance with reality. Frescoes like *Delivery of the Keys* by Pietro Perugino and the *School of Athens* by Raphael are excellent later examples of that.

We began by talking about how focusing on relative magnitudes and everyday tasks helps make a complex topic more relatable. In a way, they help create triggers in our minds that remind us to continuously think about choices. The next time you see a pile of things to be sorted, perhaps you'll think, "Ah, good and bad, fast and slow, quadratic and linearithmic." Or the other way around. Your son, daughter, niece, or nephew might ask you what binary search is, and you'll think, "Ah, freedom, William Wallace, Eppy Toam, shirts on a rack." Those kinds of associations are easy and fun to recall. Recognizing good and bad choices is aided to a great extent by knowing what the catalog of choices looks like for a particular task.

Nowadays, algorithm is a word that's on a lot of people's lips, just like big data was a few years ago; just like deep learning will be in the years to come. What I hope you'll take away from this book is that the concept is not a fad. Rather, it is deeply rooted in history, as we saw in the discussion of Babylonian tablets early on. And because it is such a timeless concept, it is worth talking about, fleshing out, and most important of all, showing how the algorithm can be used as a tool for smarter thinking.

ACKNOWLEDGMENTS

Everyone who has given this book attention has made it better. I'm indebted to Seth Fishman for bringing the project so far in such a short period of time. I'm equally indebted to Melanie Tortoroli for the book's compelling title and for her direction, insights, and edits, and to Georgina Laycock for her edits, thoughts, and suggestions. I'm grateful to Viking and John Murray for the privilege. Special thanks to my third-time collaborator, the talented Alejandro Giraldo, for his artwork, and to Sam Penrose, Elena Glassman, and Mark Reid for taking the time to review the manuscript and provide feedback on it. The link to BASIC mentioned in chapter 11 is due to Mark. For their help with work leading up to this book, I'm thankful to Mark Surman who endorsed an earlier incarnation of this project; to Peter Norvig who shared his thoughts on how to reimagine that earlier version; and to Elena Glassman who introduced me to Peter's work.

Above all, I'm thankful to my wife, Danah, and to my parents.

TO LEARN MORE

This book touched on a variety of concepts and themes. The references below expand on those concepts and give you a good overview of the theory and best practices that underlie much of what was discussed.

Ambrose, Susan A., et al. *How Learning Works: Seven Research-Based Principles for Smart Teaching.* New York: John Wiley & Sons, 2010.

> While aimed at the college level, this book contains great advice for how to help students learn. What sets the book apart is its focus on evidence-based methods and its grounding in theory.

Avirgan, Jody. "A History of Data in American Politics (Part 1): William Jennings Bryan to Barack Obama." *FiveThirtyEight,* www.fivethirtyeight.com/features/a-history-of-data-in-american-politics-part-1-william-jennings-bryan-to-barack-obama.

> There's more to politics than rhetoric. This article gives an overview of how the history of American politics can be viewed in terms of some of the concepts covered in this book, like memory, data, and links.

Bacon, Francis. *The Advancement of Learning*. London: J. M. Dent & Sons, 1984.

> There's an example in this book about tennis that always stuck with me. Here is the opening part to the quote mentioned in the preface: "So that as Tennis is a game of no use in itself, but of great use in respect it maketh a quick eye, and a body ready to put itself in all positions, so, in the Mathematics the use which is collateral, an intervenient, is no less worthy, than that which is principal and intended."

Bender, Michael A., Martin Farach-Colton, and Miguel A. Mosteiro. "Insertion Sort Is O(n log n)." *Theory of Computing Systems* 39 no.3 (2006): 391–97.

> For those interested in learning more about library sort from chapter 11.

Brooks, Frederick P. *The Mythical Man-Month: Essays on Software Engineering, Anniversary Edition*. 1975. Reprint, Reading, MA: Addison-Wesley Longman, 1995.

> A seminal book that talks about project management in the context of software engineering, with a major motif being that throwing resources in the form of people at a project doesn't always help.

Buck, Jamis, and Jacquelyn Carter. *Mazes for Programmers: Code Your Own Twisty Little Passages*. Dallas, TX: The Pragmatic Programmers, 2015.

> A comprehensive book on the algorithms behind constructing and solving mazes. I came across it near the end of 2015 when I was deciding whether to continue working on a story-based approach to teaching algorithms or abandon it in favor of something else. I was thrilled to see that some of Jamis's writings were tackling the same issues I was thinking about.

Celma, Oscar. "Music Recommendation." *Music Recommendation and Discovery: The Long Tail, Long Fail* and *Long Play in the Digital Music Space.* Berlin: Springer-Verlag, 2010, 43–85.

There is more to music discovery than the topics we covered in chapter 6. This reference is a good place to start.

Dehaene, Stanislas. *The Number Sense: How the Mind Creates Mathematics.* New York: Oxford University Press, 1999.

When I first started reading about cognitive psychology, I came across Jean Piaget (*The Origins of Intelligence in Children* . . . and so on). I then heard about Stanislas's work from an episode of *Radiolab* and was immediately intrigued by his ideas of how children's minds probably work and develop.

Demaine, Erik, and Srinivas Devadas. "6.006—Introduction to Algorithms." MIT OpenCourseWare, 2011.

These free video lectures cover algorithms in a great deal of depth and are, for the most part, easy to follow.

Dennett, Daniel C. *Intuition Pumps and Other Tools for Thinking.* New York: W. W. Norton, 2013.

I loved this book, and have fond memories of listening to it while on a solo drive from California to Florida. I would recommend, at the very least, reading the first two parts on general thinking tools.

Diagram Group. *Comparisons.* New York: St. Martin's Press, 1980.

The working title for *Bad Choices* was *Comparisons*, a hat tip to Diagram Group's title. The book lays out its catalog of facts using illustrations and scales, which I absolutely love.

Erasmus, Desiderius. *A Handbook on Good Manners for Children: De Civilitate Morum Puerilium Libellus.* Edited by Eleanor Merchant. London: Preface, 2008.

> I didn't realize until recently, having watched the BBC miniseries *Wolf Hall*, that Desiderius was a friend of Hans Holbein the Younger, who drew iconic portraits of Thomas More, Thomas Cromwell, and many others. My favorite line from Thomas More is "All I have, all I own is the ground I stand upon. That ground is Thomas More. If you want it, you must take it." My favorite line from Desiderius is "There are some who teach that a child should hold in digestive wind by clenching his buttocks. But it's not good manners to make yourself ill in your eagerness to appear polite. If you can go somewhere else, then do that on your own. But if not, as the oldest of proverbs goes, 'let him disguise the fart with a cough.'"[*]

Feynman, Richard Phillips, and Ralph Leighton. *What Do You Care What Other People Think?: Further Adventures of a Curious Character.* New York: W. W. Norton, 2001.

> These two books are an essential read for anyone interested in understanding how an accomplished scientist like Feynman looked at the world. I first read them as an introverted graduate student in Pittsburgh in 2005, and they're two of probably only a few books that I can say permanently changed my outlook on life.

Feynman, Richard Phillips, Ralph Leighton, and Edward Hutchings. *"Surely You're Joking, Mr. Feynman!": Adventures of a Curious Character.* New York: W. W. Norton, 1997.

---

[*] According to one apocryphal translation, the eminent scholar concludes with the line, "Furthermore, whomsoever denied it supplied it."

Fredricks, Jennifer A., Phyllis C. Blumenfeld, and Alison H. Paris. "School Engagement: Potential of the Concept, State of the Evidence." *Review of Educational Research* 74 no. 1 (Spring 2004): 59–109.

> I liked the discussion of engagement in this paper and what makes something engaging for a class of students.

Freire, Paulo. *Pedagogy of the Oppressed*. England: Penguin Books, 1996. First published in 1970 by The Continuum Publishing Company.

> The metaphor of the student as a vessel is inspired by Freire's "banking" model of education. The book's talk of prescription (vs. choice) as a means of oppression struck me, as did the elaborate passages about how active and critical thinking about oneself and the topics that one aims to learn are indeed tools of emancipation. This is a tremendously enlightening book.

Gefter, Amanda. "The Man Who Tried to Redeem the World with Logic." *Nautilus*, Feb. 5, 2015.

> Great ideas can come from the unlikeliest places. The essay does a great job of detailing the rise of Walter Pitts and his contributions to cognitive neuroscience. The von Neumann example in the preface is from this piece.

Gordon, Deborah M. "The Collective Wisdom of Ants." *Scientific American*, Feb. 1, 2016.

> The statement in chapter 4 about how ants may use a form of backtracking comes from this essay.

Hamilton, Edith. *Mythology: Timeless Tales of Gods and Heroes*, 1942. Boston: Little, Brown, 2012.

> The Theseus passages in chapter 4 are from this book.

Hinshaw, Drew, and Joe Parkinson. "For World's Newest Scrabble Stars, SHORT Tops SHORTER." *Wall Street Journal*, May 19, 2016.

> The example in chapter 8 about the counterintuitive approach to winning in Scrabble is from this article.

Hodges, Andrew. *Alan Turing: The Enigma*. New York: Audible Studios, 2012.

> The work done by Turing and his colleagues at Bletchley Park in the 1940s is worth reading about, as it relates to some of the concepts we touched on. For instance, the innovation that vastly reduced the time it took the Bombe to decipher the Enigma's settings was knowledge of which settings to start at.

Holt, Jim. "Numbers Guy: Are Our Brains Wired for Math?" *New Yorker*, March 3, 2008.

> Another good read about Stanislas Dehaene's work. It was the very first time I had seen the words "has a glabrous dome of a head" arranged in that order.

Knuth, Donald E. "Ancient Babylonian Algorithms." *Communications of the ACM* 15, no. 7 (1972): 671–77.

> The examples of Babylonian algorithms are from this paper.

———. *The Art of Computer Programming, Volume 1: Fundamental Algorithms*. Reading, MA: Addison-Wesley, 1973.

———. *The Art of Computer Programming, Volume 3: Sorting and Searching*. Reading, MA: Addison-Wesley, 1973.

> Though not for the fainthearted, professor Knuth's books are impressively thorough and provide a wonderful level of historical context and a great deal of mathematical rigor.

Papert, Seymour. *Mindstorms: Children, Computers, and Powerful Ideas.* 2nd ed. New York: Basic Books, 1993.

> I came to Seymour Papert interested in learning more about his ideas on constructionism—a theory inspired by Piaget's constructivism. Papert's approach advocates discovery, group, and project-based activities as a means for learning. His writings influenced the ideas that led to this book.

Poundstone, William, ed. *Labyrinths of Reason: Paradox, Puzzles, and the Frailty of Knowledge.* New York: Doubleday, 2011.

> The Stanhope example in chapter 4 comes from this book.

Sedgewick, Robert, and Kevin Wayne. *Algorithms.* 4th ed. Reading, MA: Addison Wesley, 2011.

> If you are to read one textbook on algorithms, I'd recommend it be this one. The graphical representations of many of the concepts explained in the text are a great aid to understanding how the algorithms work.

Simon, Herbert A. *The Sciences of the Artificial.* 3rd ed. Cambridge, MA: MIT Press, 1999.

> The talk about multiple outcomes in the preface is partly inspired by Herbert Simon's talk of "satisficing solutions" in this book.

Spangher, Alexander. "Building the Next *New York Times* Recommendation Engine." Open, *New York Times*, August 11, 2015.

> This blog post offers additional details on what it takes to build a recommendation engine for text-based documents.

Turing, Alan M. "Proposals for the Development in the Mathematics Division of an Automatic Computing Engine (ACE)." Report E882, Executive Committee, NPL, February 1946.

> The passage in chapter 3 comes from this paper.

Vanderbilt, Tom. "How Your Brain Decides Without You." *Nautilus,*
Nov. 6, 2014.

> The statement in chapter 8 about how we often mold new informa-
> tion to fit what we already know comes from this essay.

Wagner, Tony, and Ted Dintersmith. *Most Likely to Succeed: Preparing Our
Kids for the Innovation Era.* New York: Simon & Schuster, 2015.

> The statement about how most of the greatest contributors to civiliza-
> tion were apprentices rather than note takers comes from this book.
> It's a wonderful book on learning.

Weiss, Mark Allen. *Data Structures and Problem Solving Using Java.* 3rd
ed. Reading, MA: Addison-Wesley Longman, 2002.

> The example in chapter 10 about dominant terms comes from this
> book.

Wilson, Brent G. *Constructivist Learning Environments: Case Studies in
Instructional Design.* Englewood Cliffs, NJ: Educational Technology, 1996.

> While reading about developmental psychology, this is one of several
> texts I came across on constructivism, namely, the idea that explora-
> tion and play can aid in cognitive development.

Wing, Jeannette M. "Computational Thinking." *Communications of the
ACM* 49, no. 3 (2006): 33–35.

> As I was preparing the manuscript for this book, I stumbled upon
> this work by a former professor of mine and was thrilled to see her
> talking about a similar idea. This paper provides a different perspec-
> tive on the topic of algorithmic thinking.

RATES OF GROWTH

One of the main focuses of this book has been on comparing alternative approaches to completing the same task. In nearly every chapter, we made the comparison using a graph of rates of growth. The lines in those graphs were intentionally unlabeled. Here is a summary of those rates of growth, ordered from slowest (best) to fastest (worst).

CONSTANT TIME: Given a number of items, if we double that number of items, the time required for the task to complete will remain the same.

LOGARITHMIC TIME: For a large enough number of items, if we double the number of items, the time required for the task to complete will approximately increase by one.

LINEAR TIME: For a large enough number of items, if we double the number of items, the time required for the task to complete will approximately double.

LINEARITHMIC TIME: For a large enough number of items, if we double the number of items, the time required for the task to complete will approximately double and increase by one.

QUADRATIC TIME: For a large enough number of items, if we double the number of items, the time required for the task to complete will approximately quadruple.

EXPONENTIAL TIME: For a large enough number of items, if we increase the number of items by only one, the time required for the task to complete will approximately double! The faded-out line at the leftmost edge of each of the graphs in this book is an exponential-time line.

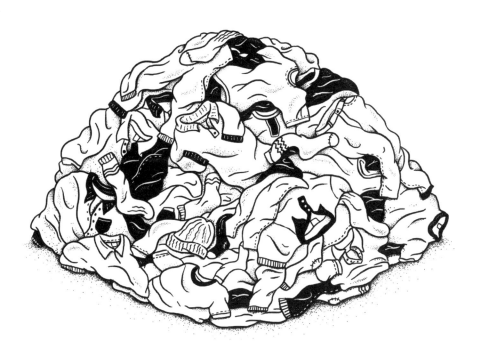

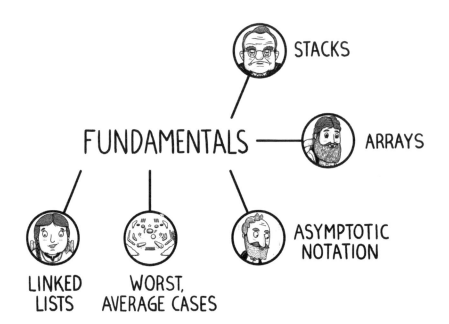

STACKS

FUNDAMENTALS — ARRAYS

LINKED LISTS

WORST, AVERAGE CASES

ASYMPTOTIC NOTATION

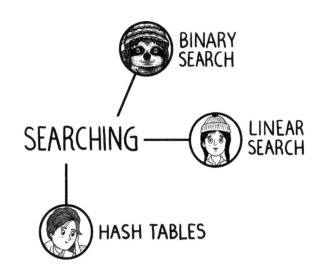

BINARY SEARCH

SEARCHING — LINEAR SEARCH

HASH TABLES

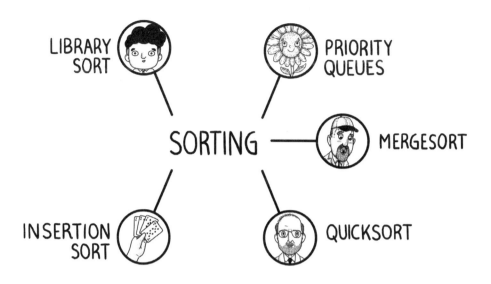

LIBRARY SORT

PRIORITY QUEUES

SORTING — MERGESORT

INSERTION SORT

QUICKSORT

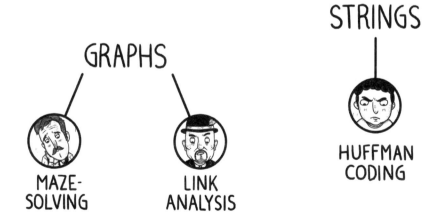

GRAPHS

STRINGS

MAZE-SOLVING

LINK ANALYSIS

HUFFMAN CODING

INDEX

INDEX

INDEX

From Byron, Austen and Darwin

to some of the most acclaimed and original contemporary writing, John Murray takes pride in bringing you powerful, prizewinning, absorbing and provocative books that will entertain you today and become the classics of tomorrow.

We put a lot of time and passion into what we publish and how we publish it, and we'd like to hear what you think.

Be part of John Murray – share your views with us at:

www.johnmurray.co.uk
johnmurraybooks
@johnmurrays
johnmurraybooks